Already published in The American Maritime Library

Master of Desolation

The Reminiscences of Capt. Joseph J. Fuller

THE AMERICAN MARITIME LIBRARY

VOLUME IX

Edited, with introduction and notes by
Briton Cooper Busch

Mystic Seaport Museum, Inc.
MYSTIC, CONNECTICUT

Master
of Desolation

The Reminiscences of
Capt. Joseph J. Fuller

Designed by Ken Diamond

Manufactured in the United States of America
First edition

Contents

Illustrations

ACKNOWLEDGMENTS

Captain Joseph J. Fuller's memoirs would not have found their way into print without the help of many friends. Bob Farwell, director of the Cold Spring Harbor Whaling Museum, originally mentioned Fuller to me. Many kind people at Mystic made my subsequent work possible: Donald Judge, former librarian, and Gerald Morris, his successor; Doug Stein, Lydia Frank, and Anne Goodrich of the manuscripts collection; and William Peterson of the curatorial department, all came to know Fuller only too well. Many other library and curatorial staff members came to my rescue more than once. In New London, Harold Cone and Elizabeth B. Knox of the New London County Historical Society, the staffs of the New London Public Library, the Cedar Grove Cemetery, and the New London *Day*, and Dale and Sharon Plumer were all most helpful. Richard Porter and Samuel Smith, Jr., of Niantic helped me trace the history of the manuscript. Captain Fuller's descendents, Janet Burgess and Ruth Sanders, provided me with much valuable family material. At greater distance, Helen Winslow of Nantucket, Richard B. Trask, town archivist of Danvers, Richard Kugler of the Whaling Museum, New Bedford, and my Colgate University colleague, Kit Hinsley, all lent their special expertise. Jeannie Kellogg and Helen Payne, who typed the manuscript, and Dean Wise, who prepared the map, were more than simply uncomplaining and efficient.

I owe a different but important debt to my co-captains, Bud Foulke of Skidmore College and Ben Labaree of the Williams-Mystic program, and to the students of Colgate, Skidmore, and Williams colleges who accompanied us on worthy voyages of the spirit aboard the *Joseph Conrad* during the January intercessions of

1976 and 1977. Without those voyages, Fuller's far more challenging exploits would still rest in the ledger pages in which he set them down.

Briton Cooper Busch
December 1978

INTRODUCTION

When Capt. Joseph Johnson Fuller died in New London, Connecticut, in 1920, the New London *Day* rightly noted the passing of one of the last of a special breed, the hardy and hard-driving captains of the nineteenth-century New London whaling fleet. Fuller was eighty when he died; he had first gone to sea before the Civil War, more than sixty years earlier. In the thirty years between 1864, when his Civil War service ended, and the return of his last voyage in 1895, Fuller spent no more than three and a half years in the United States. The remaining years were spent mainly in the southern Indian Ocean, primarily at Kerguelen Island, known to American seamen as Desolation. If any man could be said to be the "Master of Desolation," in the late nineteenth century, it was Joseph J. Fuller.

Few whaling captains, however, troubled to set down their experiences in more than the deck logs that are today so numerous in seacoast town libraries. Fuller recounted his adventures, at several points in his career, laboriously filling schoolboy copybooks or foolscap pages with his recollections of Kerguelen Island. These manuscripts, now in the G. W. Blunt White Library at Mystic Seaport, together form a unique document. Others have written of the technical side of whaling, and, to a lesser extent, of sealing, but Fuller's account concentrates on the hunting of sea elephant (elephant seal), or "elephanting" in the idiom of the day. This important aspect of the New London whaling industry is too little known, its returns hidden away in columns of figures of "whale oil" brought into port.

The island itself was first discovered in 1772 by the Breton explorer, Yves-Joseph de Kerguélen-Trémarec (1734–1797), but Capt. James Cook named it "Desolation" during his visit to the island in 1776. By the turn of the century, Nantucket and New Bedford whalers had visited Desolation Island, but British whaling vessels were the first to exploit its large colonies of seal and sea elephant. By the 1830s, British, American, and French sealers and whalers were active here, to say nothing of other visitors such as Capt. James Ross, who in 1840 extended Cook's explorations. As the fur seals were soon exterminated beyond the promise of commercial profit, the focus of their expeditions shifted almost exclusively to sea elephant (although the vessels were very willing to take any whale or seal they encountered). By the mid-1840s, the perseverance of New London masters made this island — a round trip of some 23,000 miles — very much their domain, although some other New England ports were occasionally involved. In the late 1850s, Heard Island, 250 miles to the southeast, was discovered and exploited, and for the next decade or so the trade continued as the several species were progressively exhausted. By the mid-1870s, the decline in trade was apparent; by the mid-1880s, it was nearly over.[1]

This latter stage is the subject of the second half of Fuller's book. In 1880 he sailed for Desolation in the schooner *Pilot's Bride*, convinced that he had the secret of an unexploited seal rookery. It was his misfortune, however, to be wrecked on the island, and the bulk of the manuscript describes his eleven-month struggle for survival there. Others had been shipwrecked on Kerguelen. Indeed, John Nunn's interesting *Narrative of the Wreck of the "Favorite"* . . . , published in England in 1850, describes a considerably longer stay on the island in the 1820s. But Nunn, a crewman in a party of four, was wrecked at a time when the traffic to Desolation was high, and signs of human activity on the island, so necessary to survival, were plentiful.[2] Fuller was wrecked in 1880, when the trade was nearly dead and only one or two ships a year ventured to those parts for the last sealskin or barrel of "whale oil" — and he had more than twenty men under his command.

For that reason, Fuller's narrative is a very human document, recounting the rise of a man with considerable drive from "green hand" to experienced master in a hard profession, commanding men who, despite the high level of expertise the profession required, were often the dregs of society. Not surprisingly, Fuller, in his own account, emerges successfully from his several trials, perhaps aided on occasion by hindsight. But his moments of indecision, the dangers in which he was placed by his mutinous crew, his naïveté in dealing with his ruthless superiors — these things come through between the lines. The relationships of master and crew, officers and men, were complex and sophisticated, compounded by shipboard discipline, differing socioeconomic backgrounds, conflicting levels of technical expertise, and the shared problems of simple survival. The memoirs are not only a study in group relations, however, for Fuller himself emerges as an interesting representative of his age, filled with curiosity for the natural circumstances of his Indian Ocean world, bringing home the skeletons of sea elephant or potted samples of "Desolation cabbage" in his cabin, but unaffected by the mass slaughter of animals for which he was responsible. He would have agreed with Rallier du Baty: "One need not sentimentalize over sea elephants. Their only use to the world is to provide blubber. . . ."[3]

Fuller was, in short, not wrong in believing that his memoirs should be published. He was only premature.

Fuller's background is also unique, for he was born on the remote South Atlantic island of Tristan da Cunha. The Fuller family itself had long been settled in Danvers, Massachusetts, tracing its ancestry from Sgt. (or Lt.) Thomas Fuller, who arrived in the American colonies in 1638, and on through revolutionary war veteran Timothy Fuller (c. 1707–1796), who fought, at a rather elderly age, at Lexington and Bunker Hill. Joseph's father, also named Joseph Johnson Fuller, was born in 1811 in Hudson, New Hampshire, where a branch of the family had settled. The elder Joseph ran away to sea at an early age, serving on New London whale ships and rising to the rank of third mate.[4]

During the 1820s and 1830s, the whaling grounds in the South Atlantic were the center of exploitative activity, and ships often called at Tristan da Cunha to trade potatoes and other supplies for fresh meat and produce. This island was first settled by the British with a military garrison at the time of Napoleon's exile on Saint Helena, on the theory that it might be used as a base for an attempted rescue of the former emperor. The danger was imaginary, and the troops were withdrawn in 1817, but Cpl. William Glass, who had arrived from Cape Town with a Royal Artillery detachment (bringing with him his wife and two children), and two other men received permission to stay on. Glass and his associates were joined by others, mainly survivors from wrecked vessels, and a few women were brought from Saint Helena. When Fuller arrived in the 1830s, the colony was viable and growing. Glass, honorary "governor" of the small population until his death in 1853, encouraged visiting whalemen to stay. For that reason, Fuller probably had little problem in 1833 marrying one of Glass's daughters, Mary Ann (born on Tristan in 1817).[5]

The Fullers stayed for fifteen years on Tristan. Six of their eleven children were born there, including Joseph, the author of this book. Joseph was born on 13 October 1839, according to Danvers census records, but later newspaper accounts — and his New London death certificate — give his birth date as 1840 (his tombstone in New London's Cedar Grove Cemetery simply complicates matters by giving his age at death as eighty-eight). Fortunately, such minor chronological puzzles are a rarity in Fuller's life, thanks to the excellent data available on American whaling voyages.

Fuller, senior, returned to Massachusetts in 1848, according to the account of another of his sons, Timothy, to have a lip cancer treated (there were no doctors on Tristan). Moses, four years older than Joseph, had already shipped aboard a whaler at Tristan; he will be met with serving under Joseph on the *Pilot's Bride*. Joseph himself, then ten years old, was sent to stay with his grandfather Timothy, apparently in New Hampshire. Fuller, senior, decided to remain in Danvers, sent for his wife and the rest of his children from Tristan, and became a small farmer and shoemaker (a common combination at the time), dying in 1879. The house in which the Fuller family lived, at 139 Center Street, still stands. William Glass's widow similarly left Tristan to take up residence with her children in New London when her husband died, which explains why Fuller could be rescued from Desolation by Capt. Robert Glass of New London who was at the same time his uncle, so interconnected were the few Tristan families.

Of Joseph's childhood we have only the briefest account. In August of 1859, he sailed as a "green hand" from New London on the schooner *Franklin* for Desolation. His Indian Ocean experiences were interrupted by Civil War service from 1862 to 1864 in the Union Navy as an ordinary seaman on several vessels, most notably under Admiral Farragut in the Mississippi blockade. Joseph's wartime experience seems not to have played a notable role in his life, however, and less than a month after his discharge he was headed for Desolation, as boatsteerer on the schooner *Roswell King*. Boatsteerers, roughly the equivalent of petty officers, had considerable responsibility as harpooners (when a whale was harpooned successfully, the harpooner exchanged places with the whaleboat's officer, normally a mate, who performed the actual killing of the whale; boatsteerers could, however, command the whaleboat on occasion, and often took charge of nighttime watches on the vessel itself). Promotion to this rank, as in Fuller's case, could be the stepping stone to command. Fuller became first mate in due course and sailed, four days after his marriage to Miss Jane M. Adams of New London on 25 June 1870, if the records are accurate, for a three-year voyage to Desolation as master of the *Roswell King*.

All of Fuller's known voyages, and their returns, are catalogued in appendix 1. The first half of his memoir, after describing his early life, covers the voyage of 1873 to 1875 in detail. Several other voyages followed, including the adventure on the schooner *Pilot's Bride* which is described in the second half of the book. Fuller returned from the wreck of the *Bride* in 1883 and four months later sailed again for Desolation as master of the *Francis Allyn*, the ship that had rescued him. Six more voyages followed in the same schooner, the last three from New Bedford after the vessel had been sold away from New London. When the *Allyn* returned in August of 1895, with 1,630 barrels of "whale oil," made, as usual, from sea elephants (since this schooner was just over 100 tons burden, at least two-thirds of this cargo would have had to be shipped home by other means), Fuller concluded his long career at sea.

Fuller was too active a man to retire altogether, and at the outbreak of the Spanish-American War, he obtained employment as assistant keeper of Race Rock Light (from April to November of 1898), moved to Sakonnet Light (Rhode Island), and then, in late 1899, to Stonington Lighthouse. He remained keeper at Stonington until April of 1918, living in the keeper's house, which still adjoins that lighthouse. The lighthouse itself was no longer active in Fuller's time; his responsibilities were rather the maintenance of several lights on Stonington breakwater. Nor was Fuller's career over yet, for upon his second retirement he became an elevator operator in the Harris Building in New London. The Harris Building (now the Lena Building) was and is one of the New London's more impressive structures, and in Fuller's time its hydraulic elevators were a popular attraction.[6]

Shortly before his death at the end of 1920, Fuller retired for good to his house in New London. He is buried next to his wife and two of his three children in the plot that he purchased in Cedar Grove Cemetery. Although never one of New London's merchant elite, he was remembered as an interesting and colorful man who had made his contribution to New London's seafaring history.

Although Fuller's own account gives a full picture of his adventures on Kerguelen, he says little about his employers or the firm's place in New London commerce. Until the very end of his career, Fuller sailed for C. A. Williams, a (perhaps *the*) leading New London shipowner and agent in the whaling industry during Fuller's time. C. A. Williams was the son of Maj. Thomas W. Williams, the man generally credited with developing whaling as a permanent industry in New London and one of the first to insist on temperance on his ships. T. W. Williams was in partnership with another well-known agent, Henry P. Haven. When Williams was elected to Congress in 1838, he became a silent partner, and Haven took over active control of the firm. Agents normally had a controlling share in a vessel's voyage, managed capital investment in ship and stores, signed on master and crew, and distributed profits and shares in the vessel's cargo. Such partnerships could own corporate shares in a vessel, or the partners might separately own shares in the same vessel under their own names.

After 1848, Williams returned to active membership in the firm, now Williams and Haven. During the next two decades, further partners were added, including, in 1858, Williams's son, Charles Augustus Williams (1828–1899); Richard Haven Chapell (1826–1874), raised in Haven's house (hence his adopted middle name), and Ebenezer "Rattler" Morgan (1817–1890), one of New London's most famous captains. These men had shares in the several vessels on which Fuller sailed; all contributed to the firm's prosperity, particularly in the exploitation of the Alaska fur seal colonies. C. A. Williams, in particular, served for some years as the firm's Hawaii representative.[7]

The continued growth of the enterprise was forestalled by the tragic early death at age twenty-three of Haven's promising son in 1870 and the deaths of both Chapell and T. W. Williams in 1974. Charles Augustus Williams remained Haven's partner until the latter's death in 1876. Williams, a successful entrepreneur both in the firm and upon his own account, then assumed control of the company, soon renamed C. A. Williams and Co. He was one of New London's most noted citizens, serving as mayor from 1885 to 1888, president of the New London Historical Society and of the Public Library, senior warden of the Episcopal Church, and so on. But his partners were gone, and the industry itself was in serious decline, owing to the extermination of whale, seal, and sea elephant stocks and to the development of the petroleum industry. The number of sailings for Williams declined, as they did for New London as a whole. The sale of the *Francis Allyn*, Fuller's last command, to Thomas Luce, a leading New Bedford agent, for use in the more traditional one-year Atlantic whaling voyages marked the end of more than Fuller's sailings from New London's Thames River.

In 1892, C. A. Williams and Co. was dissolved, the end of a long tradition: the Williams and Haven interests over the years had sent out more vessels by far than any other New London firm. In 1845, at the height of the industry, the partners owned eleven whalers. C. A. Williams under his own name had sent out twelve vessels, making twenty-three voyages between 1878 and 1892 — four under Fuller's command.

Williams and Haven had rivals, however, among whom the

firm of Lawrence and Co. occupied a prominent position. This company was founded by Joseph Lawrence (1788–1872), an immigrant from Italy who rose from seaman to captain to merchant prince. Lawrence entered the New London whaling business in 1833 and soon became a major operator in the trade. On his retirement in 1844, the firm was continued by his sons Francis Watson Lawrence (1821–1895) and Sebastian Duffy Lawrence (1823–1909) until 1892, although their last whaling voyage was sent out in 1887.

The Lawrences also operated in the Indian Ocean, at both Desolation and Heard islands. Before Fuller sailed in the *Pilot's Bride,* he was involved in a considerable discourse with one of the Lawrence brothers (which one is not made clear) and the captain of Lawrence's bark *Trinity.* The rivalry of the two firms, and the desire of each to steal a march upon the other, is very clear from Fuller's account. When Fuller was wrecked at Desolation, he relied upon stores that had been landed by the *Trinity;* the bark was concentrating upon Heard Island, 250 miles away, but since there was no safe anchorage there, it was customary to use Desolation as a base for operations. The end of Fuller's book recounts the legal struggle over payment for those stores, the value of which was hotly disputed. The bitterness of this conflict over six or seven hundred dollars' value (the difference between Fuller's and Lawrence's evaluation, for which see appendix 4) must be understood in an atmosphere in which every penny mattered, when a voyage of a year (formerly, two or three years had been the rule) to Desolation might well bring a return of under $10,000, and when both of the rival firms would cease operations in less than a decade.

Fuller says little about finances in his account, but it was obviously an important matter to a man who lived by the success or failure of his vessel's enterprise. As a common practice in the whaling industry, ownership in the vessels used in the dangerous waters of Desolation (normally small schooners of from one to two hundred tons, fore- and aft-rigged for ease of handling off wild coasts and for the savings in manpower such rigs offered over square-rigged vessels) was spread among a number of individuals, not merely among members of the firm. The master usually possessed a share of this ownership, although he might well have no other investments, as appears to have been the case with Fuller. The three schooners that Fuller commanded were owned as follows (sample registrations; each vessel was re-registered when shares changed hands):

Roswell King

(built at Rochester, Massachusetts, 1837; 74' length, 23 7/10' breadth, 8 9/10' depth, 96 88/100 tons net; registration of 1877–1880)

Richard H. Chapell, 5/32
Henry P. Haven, 5/32
Thomas W. Haven, 1/32
Robert Coit, 1/32
Copartners Henry P. Haven, Richard H. Chapell, C. A. Williams, and Thomas W. Haven, 4/32
Estate of Edwin Church, 2/32
Dennis Mahoney, 2/32
David W. Wetmore, 2/32
Heirs of Robert B. Minturn, 2/32
Irving Grinnell, Julia Grinnell, and Julia Grinnell Bowdin, 2/32
J. W. Frothingham and Sons, and Charles S. Baylis, copartners, 4/32
Joseph J. Fuller, 2/32

Pilot's Bride

(built at Rockland, Maine, 1856, 99' length, 26 6/10' breadth, 9 6/10' depth, 193 1/2 tons net; registration of 1880)

C. A. Williams, 7/32
H. L. Crandall, 2/32
Trustees of Estate of Henry P. Havens, 4/32
Trustees of Estate of Richard H. Chapell, 5/32
Ebenezer Morgan, 6/32
Elisha J. Chipman, 2/32
Joseph J. Fuller, 6/32

Francis Allyn

(built at Duxbury, Massachusetts, 1869; 85 4/10′ length; 22
3/10′ breadth, 8 6/10′ depth, 106 1/2 tons net; registration of
1883–1887)
    C. A. Williams, 9/32
    H. L. Crandall, 4/32
    Jane C. Allyn, 2/32
    Joshua C. Learned, 1/32
    Ebenezer Morgan, 5/32
    Robert R. Willets, 3/32
    Joseph J. Fuller, 8/32[8]

Fuller's investments may be followed in such registry records.
For example, when the *Roswell King* was re-registered in 1880,
Fuller's shares had been taken over by Lorenzo B. Chipman, the
schooner's new captain, for Fuller had transferred his investment
to the *Pilot's Bride*. With the loss of this schooner, of course, Fuller
stood to take a substantial loss, for the estimated value of the
vessel was ten thousand dollars and of her cargo another twenty-
five thousand dollars, according to the wreck report filed by C. A.
Williams and Co. The vessel was insured for nine thousand dollars, however, so Fuller was not wiped out.[9] He lost only his share
of the cargo and of the difference between estimated value and
insurance, but against this he could offset his share of the valuable
cargo already sent home from Cape Town.

    Fuller therefore had some funds to invest in the *Francis Allyn*
when he took command of her upon his return from the wreck of
the *Pilot's Bride*. He kept this share, even when the ship was sold
away to New Bedford. The *Allyn* made one voyage under another
master from 1889 to 1890, and then Fuller reached an agreement
with the New Bedford agent who was handling her, although his
share was reduced, in the re-registration of 1890, from 8/32 to
4/32. After returning in 1895, Fuller sold his shares in the *Allyn* to
the Luce interests, perhaps fortunately, for the schooner was destroyed by fire in Hudson's Bay in 1902.[10]

    The master of such a vessel had another interest, however, in the voyage: his share of the actual proceeds of the voyage, or "lay." It was also common practice in the industry to make payment to all hands by such a method, for the obvious purpose of increasing the zeal of officers and crew alike for the cause of whaling or "elephanting." A typical breakdown of such parts was that for the *Pilot's Bride* on departing New London, as given in her obligatory "Whaleman's Shipping Paper":

Joseph J. Fuller	Master	1/12
E. J. Chipman	Mate	1/16
Moses S. Fuller	Second Mate	1/32
Luke P. Gray	Third Mate & Boat-steerer	1/52
Charles A. Odell	Boat-steerer	1/70
George A. Manice	Boat-steerer	1/75
James H. Glass	Cooper, Carpenter, & Shipkeeper	1/50
John Thompson	Cook	1/125
A. G. Manwaring	Steward	1/100
Alexander Shields	Blacksmith	1/150
Edw. J. Carroll	Ordinary Seaman	1/160
Edwin B. Cole	Ordinary Seaman	1/160
John M. Edwards	Ordinary Seaman	1/160
T. M. Melrose	Ordinary Seaman	1/160
Thomas Flaherty	Ordinary Seaman	1/190
Chas. Fink	Green Hand	1/195
Timothy Reardon	Green Hand	1/195[11]

Whatever return the voyage produced was apportioned in such shares, the remainder naturally going to the owners, who had then to deduct their investment in outfitting the vessel. Fuller's return on a cargo valued at $25,000 would be approximately $2,100, but such a return would be most exceptional: only the discovery of the untapped seal rookery, which Fuller describes, justified such a high valuation of the *Pilot's Bride*'s lost cargo. A more typical example would be the 650 barrels of "whale oil" taken by the *Francis Allyn* in her voyage of 1886 to 1887, valued at $6,552 (at approximately thirty-one gallons per barrel and an average price of 32.5¢ per gallon — a particularly low price). Fuller's 1/12 "short lay" as master would have come to $546.

Return from voyage	$6,552.
Deduct crew costs (on average, 30%)	1,967.
Remaining to owners	4,585.
Deduct outfitting costs (probably $50–$75 per man per year food and supplies, excluding refitting, etc.) roughly,	1,500.
Remaining to be divided among owners	3,085.
Fuller's 8/32 owner's share	771.
Fuller's 1/12 master's lay	546.
Fuller's income for year	$1,317.

Since Fuller, as owner, would also have shared in refitting costs, no allowance has been made for his offsetting probable share of profits from the ship's stores or "slop chest." An estimate of his general yearly income at between thirteen and fifteen hundred dollars on average would not be far off: in the 1880s certainly an acceptable income, but not enough to qualify Fuller as an eminently wealthy man, able to buy into the firm in the style of a "Rattler" Morgan.[12]

By comparison, at the other extreme, an experienced hand's "long lay" of 1/160 (again, fairly low for a small ship on a one-year voyage) came to a mere $41 for the same cargo. If in the course of the year's voyage he had drawn clothing, tobacco, or other supplies from the vessel's slop chest and an outfit or cash advance before sailing (for which he would be certain to pay interest, often 25 percent per annum), in addition to sundry charges for expenses for the medicine chest, or for ship loading and unloading in harbor (in theory, justified because the crewman, dumped aboard shortly before sailing and departing when the last sail was furled, was unable to participate in loading or unloading), he might well finish his voyage in debt to the owners as a result of his year's work. It was a sweated industry, and for the men concerned, the industry was "at its best, hard, and at its worst represented perhaps the lowest condition to which free American labor has ever fallen."[13] By such standards, Fuller was indeed a rich and successful man.

The object of Fuller's endeavors and the source of his livelihood, Kerguelen Island, is extensively described in his account from a geographical standpoint. Similarly, he has much to say about its flora and fauna, both in the text and in his own additional notes (appendix 3). Nowhere, however, does he specifically comment upon its climatic conditions, although no reader can escape the endless sequence of squalls, gales, rain, and wind as Fuller waits impatiently for the rare calm days.

Kerguelen's trying climate is due to its position just north of the fiftieth parallel, the zone of convergence of layers of cold currents circling out in a northeasterly direction from the Antarctic that meet the warmer currents sweeping down from Madagascar and the Cape and are assisted in their counterclockwise passage by the strong westerly winds — the "roaring forties" and "screeching fifties" — winds averaging twenty or twenty-five knots over the year at Kerguelen, with maximum ranges to seventy knots or more.

Where the polar water is driven beneath the lighter, warmer air masses from the north, considerable turbulence and precipitation result, with much fog and low cloud cover: the name Cook gave the Cloudy Islands, just to the northwest of Kerguelen, was no accident. Incoming masses of polar air meeting warmer air from the "horse latitudes" form intense low pressure systems, which move steadily to the southeast (or east, if they have formed in the South Atlantic off Argentina). Kerguelen, as a result, experiences a very constant west wind, with heavy breakers, waves rolling across thousands of unobstructed miles — at all times of year, 15 percent or more of the waves will be over twenty feet high (the tidal rise, on the other hand, is under five feet). The combination of wind, sea, and a long jagged line of rocks and high cliffs makes Kerguelen's west coast one of the least attractive lee shores in the world.

Bligh's Cap

Cloudy Islands

Tony's Harbor

Terror Reef

Swain Islands

Rocky Bay

Table Mt.

Cape Français

Christmas Harbor

Soulskin Bay

Mt. Havergal

Arch Point

Foul Hawse Bay

Muscle Bay

Cumberland Bay

Breakwater Bay (Cassell Cove)

Howe Is.

Port Fuller (Little's Harbor

Saddle Hill

Mt. Palliser

Port Cook

Bee Hives

Center Bay

White Bay

Port Palliser

Penguin (

Eclipse

Bear Up Bay

Rhodes Bay

Hopeful

Black (Ty.
Bay

African Bay

Whale Bay

Hillsb
B

Winter Harbor

COOK
GLACIER

Irish Bay

Thunder Harbor Bay

Shoe Foot Beach

Monument Cove

Duncan Cove

West Is.

Blueskin Beach

Cape Louis

Marianne Strait

Mussel Bay

Young Williams Bay

Bull Beach

Melissas Bay

Hot Springs

Table Mount

Table Bay

Volage Bay

Shoalwater Bay

Fortune Islands

Hell's Gate

Iceberg Bay

Boat Bay

Mt. Ross

Sprightly Bay

Bonfire Beach

Big Half Moon Beach

Cape Bourbon

Cape Dauphin

White Ash Beach

Little Half-Moon Beach

Swain's

KERGUELEN ISLANDS

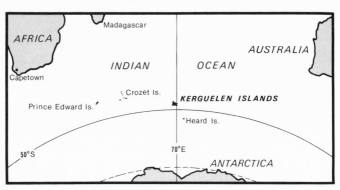

miles

D. Wise

0   5   10

elevations
in feet

Madagascar

AFRICA

INDIAN   OCEAN

AUSTRALIA

Capetown

Crozet Is.

KERGUELEN ISLANDS

Prince Edward Is.

Heard Is.

50°S

70°E

ANTARCTICA

Kent Islands

Rocks of
Despair

Blackfish Bight

Abbot Is.

Norton's Harbor

Accessible
Bay

Mt. Campbell

Cape Digby

een Is.

1378   787

Betsy Cove
(Pot Harbor)

Lake
Marville

Green Hill Royal Bay

Bay

1827

Black Point

Cape Sandwich

2395

2398

Cascade (Cold
Springs) Bay

Mt. Peeper

PRESQU'ILE   COURBET

Malloy Point

Shoalwater Bay

Observation Bay

Prince of Wales
Foreland

Hog Is.

North Is.

Somerset Pt.

Cat Is.

Three Is. Harbor

Royal

Grave
Is.

Sound

Kerguelen Head

Greenland Bay

Cape George

Cape Challenger

N

W   E

S

Climatic conditions may be summarized as follows:

Month	mean air temp. °F.	relative humidity %	cloud cover %	rainfall average inches	mean wind velocity (knots)	% winds from SW	W	NW	Calm
January	43.5	81	74	3.0	19.1	7	40	40	1
February	43.2	79	68	1.5	25.2	5	40	49	0
March	40.6	—	—	2.9	—	—	—	—	—
April	38.1	—	—	1.5	—	—	—	—	—
May	36.5	73	—	2.0	26.0	3	85	4	2
June	35.1	75	61	4.1	23.0	3	79	4	3
July	34.0	77	—	2.5	28.7	2	86	2	3
August	33.6	—	—	5.7	—	—	—	—	—
September	33.6	—	—	2.4	—	—	—	—	—
October	35.1	—	—	2.8	—	—	—	—	—
November	38.7	81	77	2.7	13.0	36	29	13	1
December	41.5	77	76	2.5	20.0	9	29	45	1[14]

The island itself — actually a group of hundreds of islands — is of volcanic origin. Fuller describes volcanic activity in
the southwestern corner of the island, particularly hot springs.
The whole is cut into innumerable peninsulas by fiords, long
narrow valleys cutting off the higher ridges and tablelands, often
headed by glacial ice, particularly in the west. Much of the main
island mass is covered with ice, formed by the persistent rain and
snow on the higher elevations, but, from a mariner's standpoint,
possessing the offsetting advantage of providing fresh water in
melting streams and rivulets. The island, on the other hand, is
surprisingly warm (again, due to the convergence), with an annual
average temperature of 37.8° F. and a range of seldom more than
10° in either direction, even in (Antarctic) winter. Heard Island,
by contrast, which lies along the same submarine ridge some 250
miles away, is much more glacial, with overhanging ice ridges that
are undercut by the sea. The difference is caused by Kerguelen's
nearness to warmer water, and as a result its conditions more
nearly resemble those of the Crozet Islands, which lie in the same
latitude.

The climatic features, however, explain Fuller's presence at
Kerguelen, for the sub-Antarctic zone (that is, the area between

the subtropical and the Antarctic zones) provides the reasonable temperatures and rich nutrients welling up from the cold water of the south that support a rich bird and marine population — including the seal, sea elephant, and whales for which Fuller searched.

A final word about the manuscript itself. Fuller wrote the first half in a copybook, beginning, "I had been attending school all winter." To this autobiographical account, he subsequently added his notes on Kerguelen as an introduction (reproduced as appendix 3). Study of the manuscript reveals that these pages were sewn in at a later date; their slightly smaller size bears witness to their separate origin.

When the manuscript was written is a moot point. Inside the copybook cover Fuller offered his own title: "Four Voyages to Kerguellan's Island. Or life in the land of Grease. Being Extracts from the Journal of Capt. Jos. J. Fuller. written by himself. '91." Unfortunately, this proves little. Internal evidence, such as the reference to passing Bermuda, indicate that Part I was written during his return voyage in 1873. Half-way through the manuscript, the ink becomes very faint, and Fuller explains in a note that he has run out of ink and has made recourse to his medicine chest for a substitute (with little success, it may be added). Since the account is quite complete, the reference to "extracts" may be regarded as a contemporaneous literary convention.

Part II, covering the wreck of the *Pilot's Bride*, can only be dated to the years after his return in 1883. Alas for consistency, at the end of Part I Fuller makes reference to a second work in progress, meaning Part II, on his shipwreck at Desolation. A nice solution would be that Fuller copied an early original draft into his copybook in 1891, adding a reference to his later disaster — but no need, then, to write in ink made of iodine. There are far too many external proofs of Fuller's career to doubt the authenticity and accuracy of the manuscripts, but the mystery of dating remains unresolved.

The later history of the notebooks is clearer. Fuller hoped for publication, and perhaps even discussed the matter with prospective publishers (he even thanks his "publisher" for assistance in his own "author's introduction"). Any such negotiations fell through, however, and the manuscripts never found their way into print in Fuller's lifetime. On Fuller's death, they became the property of his youngest child, Gertrude (1880–1969), who also inherited two watercolor portraits of the *Pilot's Bride* painted at Cape Town, Fuller's sextant, and other personal articles. Most of these items, including the manuscripts, she gave to a New London area resident, who sold them to a commercial dealer from whom they were purchased by Mystic Seaport in 1971.

As far as possible in this book, I have adhered to the original text. In some places, however, edges have frayed away, Fuller has written over an earlier passage, or the writing is simply not decipherable. Such instances, together with blanks left by Fuller for later research (normally for a particular latitude or longitude or distance in miles), presumably from logbooks or other sources, are indicated in the notes.

Fuller's style requires little alteration, but some habits, merely quaint in a paragraph or two, become grating when carried over hundreds of pages. In both manuscripts, Fuller began with good intentions in the matter of paragraphing, punctuation, and the like, but soon the text becomes one long paragraph, a series of sentences linked by the conjunction "and," rolling in unceasing repetition like the breakers on Desolation. Therefore, paragraphing and other necessary punctuation have been added. Similarly, Fuller normally jumped from tense to tense, and endings have been added for conformity of tense and gender where necessary for clarity. Fuller had disconcerting spelling and vocabulary habits as well: "where" for "were," "there" for "their," "to" for "too," "proceeding" for "succeeding," and *vice versa* in each case. Such usages, as well as modern compounding of words, have been corrected without special indications to that effect. On the other hand, words added by the editor are always indicated by brackets. Place names present a particular problem, as few used by Fuller appear on modern charts of Kerguelen. All places that could be located have been included on the map of Kerguelen; those that could not be found are so identified in the notes. Both locations and chronology have been confirmed where possible through the use of available logbooks of ships encountered and published accounts.

Fuller apologized in his manuscript for his lack of literary ability. There was no need to do so, for his adventurous life as he related it surely holds fascination. He would not have minded small editorial corrections, however; Fuller wished for such assistance, but, unfortunately, seems not to have received it in his own lifetime. Fuller's memoirs need no apology, but that, as Fuller might have said, is for the kind reader to discover.

Master of Desolation

The Reminiscences of Capt. Joseph J. Fuller

AUTHOR'S INTRODUCTION

To those that will evince any interest in this work while perusing its pages: the occurrences that are set forth in this book are what actually came to pass during four consecutive cruises to the South Seas. There has not been the slightest exaggeration made. Being homeward bound from Desolation Island and not having any thing in particular to occupy myself with, I have jotted down at intervals the happenings of those four voyages. I think that the work will receive a kindly reception from a certain class of folk, and I hope that it will be read with attention and interest by them. I assure you kind friends that every thing herein stated in regard to the part and parts that the work treats on comes from the very best source and can be considered authoritative. There are several subjects that I contradict certain gentlemen that have attempted to journalize these parts. It is not just for the sake of arguing the point that I have done it but mainly to remove false impressions. I hope that nobody will be backward in patronizing the work that is interested in the various subjects that it treats on. I can not say more but thank the Publishers who has given every encouragement, and thank the multitude, I hope that will be pleased to read it.

Yrs. truly, Capt. Jos. J. Fuller

PART I

The Voyage of the *Roswell King*, 1873–1875

I HAD BEEN ATTENDING SCHOOL ALL WINTER. ONE DAY I
returned home feeling quite sick. I entered the house, put
away my books, and sat down by the fire, my head resting on my
knees. I then fell into a semiconscious slumber. I was awakened
by my father coming into the room, who upon seeing me in the
position that I was in, accosted me thus, "Joseph, what ails you?"
At these words I came to myself again, and lifted up my head.
"You seem to be ill," continued he. "I am not feeling hunky,
father," replied I. He then asked me if I had a fever. He could see
from my haggard appearance, that something was the matter with
me. I told him yes, that I felt a bit bad. He then went into the adjoining room and summoned my elder brother, to go in quest of
the doctor. In due time the doctor arrived. Little did I think at the
time, that I had been stricken down with a very dangerous if not
fatal malady. Typhoid fever had been raging in our vicinity for
the past fortnight; and in fact it had turned into a regular
epidemic. There had already been some talk of having school
closed for the season, on account of many having already suc-
cumbed to its fatal effects.

The doctor having arrived, he set about propounding questions to me: how I felt, where the ache was, etc. All of these questions I answered satisfactorily. He stood eyeing me for a moment;
he then shook his head and said, "Young man, you had better be
off to bed. And I am afraid that a good many days will elapse before you will be able to go out into the open air." I assure you that
this was not any nice news to me, as I had a picnic to attend that
week, and a game of ball to play next day. I could see that there
was no use of grumbling. Fate was against me, so I was marched
off to bed. My mother fixed up a decoction for me and after having administered it, she bade me good night and retired. I was left
to my own thoughts. Towards morning I commenced to grow

very delirious, the fever was quick in taking effect upon me. I suffered very much during the [succeeding] week; at times it was
thought that I would give up the ghost. But being very strongly
constituted I survived. Two months and some days was the dura-
tion of my sickness.

As I have already said, I had a strong and robust constitution.
But the malady had not left me without leaving its evil effects
upon me. From a strong and robust boy, I had dwindled down to
a mere skeleton of my former self. The doctor was still attending
me, but I did not seem to be improving in health a particle. Many
of my friends prescribed a change of climate for me, while others
said that a sea voyage would prove very beneficial. The latter I
preferred, as I had always had an eager desire to go to sea. There
seemed to be something fascinating about a "Jack Tar" life. And
Oh!, for adventure and romance. Like many other young men, I
had read tales of seafaring life, stories of wrecks and piratical tales.
"A life on the ocean wave" sounded quite romantic to me. One
morning after breakfast, my father asked me if I would not like to
go out into the country, and kind of recuperate my health. I re-
sponded [to] him by saying "No" emphatically. "No!" My father
was rather abashed at this.

"Well," he said, "my son, it is for you to choose; you can either
go to sea, go into the country, or stay at home. I will give you
until tomorrow to decide. You must remember before you take
this step, that a seafaring life is not one of luxury, it is one of peril
and deprivation. Your future career will depend upon the one you
choose. You are exposed to very inclement weather. I myself have
experienced it all; so you see that I don't speak from a mere
standpoint of view but from actual experience. And furthermore
it is a place where you won't learn what I desire you to, your daily
associates will be a rude uncultured lot of men, but men I dare say
kindhearted and as a general rule with principle; remember you
are making your own bed and in it you will have to lie, you won't
have no Mama or Father to look after you, and the living is very
frugal but substantial."

After this harangue, as I termed it — God forgive me for calling it so — I went out to tell my companions of it, and I was fully

determined to go to sea. I was highly elated over it I must say. My mother was frantic over it. "To make a long story short," as the saying is, it had been decided and arrangements were to [be] made to send me off to sea. I was to sail in a week's time. My father had wrote to some shipowners in New London, and had been accordingly advised. During the proceeding week, I was kept busy making arrangements for my timely departure. My mother was laying me in a stock of thick underwear, oil skins, sea boots, and in fact all the requisites that were included in a "sailors' outfit."

The eventful day at last arrived when I was to take my departure; with tears in my eyes and a big lump in my throat, I bade farewell to my father, mother, sisters, and brothers. My luggage had been sent to the depot, so I made my way to the station to take the coach. In those days we had not the facilities for travelling that we have now-a-days. The only railway around our place was a small branch that run into an adjoining town. Travelling by stage coach was very near as quick as by rail. Having arrived at the depot, the coach was then taking on passengers, express matter and mail. I mounted the coach, pushed myself in the most convenient empty seat with an air of importance, trying at the same time to give the passengers the impression that I was accustomed to travelling on such vehicles. My father and elder brother had accompanied me, and also many of my youthful companions went there to bid me good-bye.

I had got through the form of shaking hands, when I was about to settle myself down in my seat again, when I heard a familiar voice behind me. I looked around and there standing in the post office door was the post master, a Mr. Prentice.

"Joseph," he said, "you are not going off without telling me good-bye, are you?"

"Why no, Mr. Prentice," said I, "you will excuse me for being so thought[less] in not remembering you." I jumped down off my elevated position, he grasping me by the hand when I landed on "terra firma" — and these are the words he said to me.

"Joseph, my boy, you are about to go to sea, I learn. Now let me give you a few words of advice. No doubt but what your people have done so already, but a few words from me won't be amiss. First of all, let me caution you in regard to your behavior when aboard ships. At all times be obedient, respect your superiors, and your self too," he added. "Be willing and accommodating, let them know that you desire to learn, be civil to everybody, as it pays to be I assure you, and take care who you associate with, and I am confident that if you follow out these rules you will get along all right, but if you don't, I assure you that you will rue it. You yourself know whereof I speak, not merely from what I have read and heard say but from actual experience; you know the step that you are about to take, so be cautious. Now a few words in regard to money matters. I will repeat the old adage to you, 'Take care of the cents and the dollars will take care of themselves.' It is an old one, but a true one. No doubt but what you have often heard me narrate some of my exploits, of when I served on a privateer, in the War of 1812." Yes, thought I, to myself, I had. It was my special duty to go to the Post Office and get the mail, and oftimes I would tarry there to hear the recital of one of his yarns, but it often proved to my own sorrow, for generally I would be tardy in getting back home. I would find mother waiting impatiently for me. She would give me a good scolding and often threatened me with the birch.

"Well," continued the venerable looking old post master, "I don't wish to speak to you of fabulous sums of money, but to tell you the actual truth, we took enough money with the ship during the war to fill this building full of money." At the same time he pointed at the building over his shoulder, a very large structure indeed.

"And what have I now to show for it? Of course you will remember, that I came in for a round sum of this prize money. I lived rather too extravagant in my younger days. I was what you call a spend thief. Now I have but a roof over my head and but can manage to eke out an existence, whereas I should be retired and living a life of peace and comfort. 'Let your great study be in future years economy.' Now good-bye Joseph, my lad, and may Dame Fortune smile on you."

With these last words he gave my hand a hearty shake. My father embraced me and I once more took leave of my youthful

schoolmates, and took up my position on the coach. The word
was given to start, and away we went, bound for New London. I
kept looking back at my friends and I could see that some of them
cast me an envious glance, wishing that they were in my position.
It took us but a short while to get out of sight of the town. The
drive was a long and tedious one; we made a good many stops to
let off and take on passengers, mail, and express matter. We at last
arrived at New London. My uncle was waiting for me, as he had
been informed of my determination to go to sea. He greeted me
very cordially, seen that my luggage was conveyed to the ship,
and asked me up to the house. Next morning found me at the office of the agents, Mess. Williams Haven & Co. My uncle, of
course, was with me; he was to give me an introduction to the firm
and see that I was put aboard of the ship. I was received very cordially.

And now came the business part of the affair. Of course I went
through the regular form of shipping. The ship's articles were
read to me to this effect.[1] The voyage was to last two years. We
were to go in quest of elephants and whales. I shipped as a "green
hand," I having never been to sea before. My navigating had been
limited to going a'sailing on the river that flowed by my native
town. It was now August 1859 and we [were] to be back in 1861
time. I was then told to sign. It was [with] a certain amount of
boyish pride that I took pen in hand to do it. Already though I had
taken the most important step towards being a sailor. From these I
was taken aboard. My uncle introduced me to the captain — the
old man they called him aboard, he seemed to be a very nice sort
of fellow, only a little gruff to his men.[2] My luggage was taken
forward into the forecastle and a bunk assigned to me. We were to
sail that evening, if the wind came fair, as in those days tugs,
steam ones, were not so plentiful as at the present time.

At last the wind came on from the right quarter. All hands
were mustered aft and the two "watches" were chosen. Eight bells
were rung. The starboard watch went below, while the larboard
watch was busy getting things ready for sailing. We then were
served with our dinner. The fare was very frugal but still substantial. And now commenced the regular routine of a seafaring life.

After having got through we went on deck to relieve the other watch so they could go below and whet their appetites. Before going on deck I donned my working clothes, and went on deck looking like a little sailor, as my mates told me. The other watch had finished their repast, and all hands were ordered on deck to get the ship underway. We set about letting loose our moorings. The wind kept freshening up and the order was given to set sail. All sail was crowded on, and away we went skimming over the water. Getting fairly under way, our watch was sent below. The breeze was getting stronger, and of course the sea got more ragged.

We were nearing the terminus of the sound when I felt a horrible feeling coming over me! Oh! I felt so bad. I tried my best to over come it, but no, it would not work. All of a sudden, I felt a curious feeling in the region of my stomach. I then knew what was coming. I made haste up on deck, feeling more dead than alive. I went to the lee side of the ship and commenced to throw up. I thought my very entrails would come up. Of all the nasty, nauseating, and languid feelings that man is heir to, I honestly believe that seasickness caps them all. After finishing throwing up all I could, I felt a bit relieved, but only for a time, as I was at it again. Seven bells were struck and the watch called. Now to add more to my sufferings a squall was coming up in the distance.

At eight bells all hands were made to stay on deck as it was thought that we would be obliged to shorten sail. The squall was very near upon us. At last came the command, "Close reefs in your fore and aft. First the fore topsail and flying jib." I felt rather excited as all of this was being sang out by the captain. Already I had commenced to regret that I had come to sea. But those words of my father still rung in my ear. "So you make your bed so you lay in it." I assure you that a "life on the ocean wave" did not sound as nice as formerly, and in fact I felt like as if [I] could kill the man that composed the song. He surely could not have experienced what I was then going through. I was ordered aloft to help furl the topsail. I staggered along the best I could; it would be some time before I would get my sea legs. As I had never been aloft before, I was inclined to be afraid. Just picture yourself, kind

reader, in my predicament. I managed to get up on to the shear
pole, and then I started up the rigging. Rattling [that is, ratline]
after rattling I climbed up until I reached the top. It was the first
time that I had ever been aloft. So you can imagine how I felt. I
was terror stricken. When I arrived at my destination aloft, I took
a glance below, and the sight that presented itself to me made me
tremble like a leaf. The ship was rolling to and fro; the sails were
flapping like so many horrid vultures; the rigging screeching in its
fiendish glee. In all it was a trying and desperate moment for me.
Large volumes of water were constantly coming over the sides and
in fact it looked at times as if the vessel was completely submerged
under the water. It appeared to me like as if it was some terrific
storm coming on but which in reality was only an ordinary squall.

I performed my duty aloft to the best of my ability. I arrived
down as far as the fore-top, and I again felt this horrid feeling
creeping over me. Making myself secure by grasping a hold of a
topmast backstay, I tried to throw up; but in this I was disappointed; I had nothing in my stomach but my entrails, and those I
could not well part with. At last I got a spell of relief. I then
started to descend. Having alighted on deck all right, I went forward, and laid down on the fo'castle scuttle, when the mate came
along and roused me out of it. He told me to keep a moving or I
would never get over my sickness. He then said, "Why don't you
take a piece of pork, tie a rope yarn on to it, get some molasses,
and dip the pork in it, and haul it up and down your throat?" I
could not stand it any longer. I rushed to the side and tried to
throw up! But, no, it would not wash. Just the thought of anything greasy would make me feel sick. Of course this caused a
good deal of laughter, but all at my expense. For a good many
weeks afterwards I suffered at times from it. Every time that I
would get a whiff of bilge water it would turn my stomach. And,
oh, the horrid [smell] that would ascend from the galley; especially when the salt junk was being cooked, it was simply terrible.

That evening it calmed down, so I did not feel so bad. The
squall having now spent its fury, we proceeded to put on sail. The
reefs were accordingly shook out of our sails, the sails that had
been furled set again, and we went sailing along. We at last ar-

rived at the terminus of the sound. It was now fairly well along in
the night. Montauk, with its bright light shone out in the dark
night on our starboard hand and Block Island on our larboard. We
at last arrived into the ocean. With our ship heading southeast, we
sailed away from the American coast. After everything was set
a'working in ship-shape style, I commenced to take in my surroundings and to get acquainted with my shipmates. "What's
your name?" "Where do you hail from?" were the two prevailing
questions. There was one of the foremast hands who bore the
same [name] as what I did. In fact he was a "towney" of mine.
This being the fact, we were ever afterwards sworn friends. Nothing of any interest transpired aboard, only a light squall now and
again. We passed a few ships.

On the third day the horizon commenced to blacken. I was at
the wheel; in the past three days I had learnt to take my trick at
the wheel. I got so that I could steer a course as good as any of the
old "sea dogs." "Haul aft the main sheet," was sang out by the
second mate. The men came aft and done as they was bid. The
captain and first mate had been below to their dinner. When the
mate came up on deck, he glanced around and said something to
the second mate about bad looking indications to windward. The
captain came up on deck and remarked to the mate something
about the barometer falling very fast.

"You had better take in all light sail, and put in reefs in the fore
and aft sails," said the captain. His orders were obeyed in quick
time; and we went running along with short canvas. The wind
commenced to come on stronger, and the sea kept getting higher.
"I think we are going to have a regular snoozer," said the first
mate, a typical old sea dog.

"Yes, I think so myself," replied the captain. "Well, if it comes
on any stronger we will heave her to."

Pardon the interruption, kind reader, as I do not think many
of you are familiar with this nautical phrase, "heave her to." The
reasons assigned for heaving a vessel to, are these. When you get
the wind from any quarter contrary to it being fair, and if there is
a big sea on, you are obliged to take in all sail but the foresail and
that you have to close reef. You then set your storm trysail. This

sail serves to keep her — the ship — on the wind, it does not let her fall off too much and the foresail keeps her away from the wind. The combined actions of these two sails keep the ship from getting into the trough of the sea.

We at length hove to. And it commenced to blow — "green devils" — I heard one of the old "salts" remark. She rode the sea like a duck. The sea was running mountains high. It was a hard matter to get along the deck. Even those who had practiced sea legs were unable to get along without grabbing a hold of something for support. We laid to for two days. At last the wind commenced to lull; the sea was not so high as it had been. The order was given for us to take the reefs out of our foresail and take in our storm trysails. This having been accomplished, we kept her off. Towards noon we set all sail, and went sailing our course.

Nothing of any interest transpired until we arrived off of the Cape de Verde Islands. We could see Bravo [Brava] and Fugo [Fogo] looming up in the distance. We [were] to windward of them and we ran in pretty close to them. We got near enough to discern objects ashore. They looked very green and verdant. This group of islands lie in ———[3] and are under Portuguese rule. But a few hours elapsed before they grew dim in the distance. The next land we sighted was the Island of Trinidad,[4] a desolate looking piece of ground, not as inviting as the Cape de Verde Islands. No doubt but what many of my friends have read of the fabulous sums of money, diamonds, and precious jewels that are supposed to have been concealed on this island by the buccaneers during the time of piracy. It looked like a fit place for booty to be stowed away on. It had the appearance of a desolate and arid island. It is uninhabited except by sea birds and enormous big land crabs. This island lies in 20°31′ South Latitude and 20°10′ West Longitude.

Nine days after passing Trinidad Island we came in sight of the Islands of Tristan de Acunha [that is, da Cunha]. These islands lie in Latitude 37°0′ South and Longitude 12°02′ West. Inaccessible and Nightingale are the smallest of the group. Tristan de Acunha is the largest. It has the appearance, when at a distance — say a couple of leagues — of a large stone projecting out of the water. But upon approaching it you notice the low-lands, upon which a small settlement [is found]. The population of this island is about sixty. The colony is composed of ten houses. They [the settlers] gain a livelihood by trading with vessels and cultivating vegetables and raising cattle. We did not stop at this island but sailed on.

Some days afterwards we got into a school of black fish, an animal allied to the whale, but not quite so large.[5] We received orders to lower away for them. Both boats were put afloat. We paddled out towards where the school was; after a good deal of maneuvering our mate managed to get a hold of one. It was quite a novelty to me. Now had commenced the romantic part of the business. Fathom after fathom of the line was run out, as he sounded immediately upon being struck; he came to the top again, and we were engaged upon taking in the slack of the line. We managed to [get] close enough to him so as the other lance could be put into him. After a little more hauling and pulling about the monster — as I thought it was, although it was not much larger than flukes of a good sized whale — came to the top again and expired.

We towed our prize back to the schooner. After cutting his blubber off and taking what ever we wanted of him, we set her on her course again. When the blubber had been tryed out, it netted us three barrels of oil. Sunset found us many miles away from these islands. We experienced a good deal of foul weather during the [succeeding] five days. At last we came in sight of Kerguelen Island, better known as Desolation Island. This island was discovered by Captain Kerguelen in the year of 1772. It was visited by Captain Cook in the year of 1776.

At about 10 o'clock we came in full view of the island. I thought it a very hard looking country. I assure you that we found it to be the truth before getting away from it. There was a pretty stiff breeze a blowing when we started to beat up to it. We were coming in close quarters, and then I could get a good view of the land. For two or three miles along the coast, back of the lowlands there was a descent, almost abrupt enough to deserve the name of a precipice. The wind would come down off of this highly elevated land, right on to the surf, and made it look as white as a feather. Our schooner being close hauled on the wind, it would send her over her beam ends, then there would be a lull but only to be repeated again and again. To make things worse, it came on to rain in torrents. After an enormous lot of work we managed to get into Sprightly Bay. Little did I think at the time, kind reader, that I was destined to spend the best part of my life on and in the vicinity of this island. But as the old maxim runs, "God proposes and God disposes" *(sic)*.

It was high noon when we got our anchors down. As I have said before it had the appearance of a very desolate island. The land in certain positions was very elevated. A certain amount of vegetation seemed to exist on it. Here and there wheeling in the air could be seen vast multitudes of sea birds, uttering their mournful cries, while on the rugged and uninviting beach could be seen the sea elephant basking in the mid-day sun. These were the first sights permitted to view. These sea elephants, I think, are a species of the walrus — an Antarctic walrus you might call them — at least I think they are allied to that family of animals.[6] An anchor watch was set that night. We were to go elephanting next day.

That night I dreamt about them. I was dreaming that I was pursued by one; he was making a grab for me, he had got a hold of [me]. I was just waiting to hear my bones crushed between his mighty jaws, and I awoke and found my chum had a hold of me and was shaking me up and asking me what I was making all that infernal racket about. The noise that I made in my dream when I was trying to escape my terrible foe awoke him. Of course he came over to investigate the matter and he found me uttering some unintelligible words and all mixed up in the meshes of my bed clothes! By gosh, thought I, if those elephants are as bad as they were pictured to me in my dream, you won't catch me fooling around them.

After getting my course fixed in my bunk, I laid down and went to sleep until next morning without seeing anything more of elephants, my evil genius. As I did not rest good through the night I felt rather fatigued. In fact all hands presented a haggard

appearance, most [not] being in the custom of turning in for all
night since leaving port; it did not agree with them. We were kept
busy until breakfast making preparations to go ashore and slaughter the elephant. Knives, lances were being sharpened: the knives
to skin with and the lances to kill them with. Also a few of the "after gang"[7] had fire arms to shoot the critters with. Our dinner was
packed away in the boats, as we were not to come back until the
evening. At last came the order, "Lower away your boats." Both
boats' crews sprang into their boats, and we then pulled for the
shore. We had no difficulty in getting a place to land. After landing we started towards where the elephants were. We had previ-
ously received instructions of how to kill and skin them. Of course
the experienced hands were to lead the way.

Generally the way the work is carried on is thus. The officers
and boatsteerers do the killing and skinning, and the foremast
hands collect [the blubber] and carry it to the most convenient
spots so as it can be conveyed aboard. On the beach could be seen
large quantities of them. They seemed to congregate into mud
holes in great numbers. Arriving on our field of operation, the officers and boatsteerers commenced to kill them. The first and second mate were to do the shooting and the boatsteerers the lancing.
The elephant, after being shot and killed, has to be bled. They
used Winchester rifles, caliber 44. The lances are about three feet
long, and they are fixed with a socket into which fits a pole about
six feet in length.[8] You have to be cautious in killing the elephants;
don't get too near them for they will give a sudden spring towards
you and if you have no way of retreating you are gone up the
spout, for if they once get you, or any part of your anatomy, between their powerful jaws nothing will save you. At times when
you are trying to lance the critters, they will grasp the lance between their teeth, and if you keep ahold of it — like as if by
magic — you will pick yourself up, about twenty feet distant, and
when you come to look at your lance, you will find the tip of it
gone. The cows are more dangerous than the bulls, they being
more active, but not so powerful.

It takes four men to kill a good sized bull. First of all you have
to get them out of the water holes. This is accomplished by driv-

ing them out with a stick. You then commence to slaughter bulls first. The reason for doing this is because the cows will remain on the beach to protect their young, but the bulls will make straight way for the water, and nothing but a bullet in the brain or a well-delivered stab will stop them. Sometimes it takes as high as ten bullets to stop them and then a good spearing to kill them. As the elephants do not go over two or three hundred feet from the water's edge, they have not far to travel to get back. After all of the bulls have been slaughtered, the cows are killed and also the young. Those that are small are let go free. Now comes the skinning. It takes four men to skin a bull. So as the reader can form some idea of how the skinning is done, I will describe the animal to him.

No doubt you have seen pictures of seals and walruses. Well, these elephants resemble them very much, in fact they are a species of the seal. They commence skinning them by cutting them from the back of the head to their tail. Then the back is skinned down as far as the stomach. After skinning him, you commence to operate on the blubber. You cut through it until you come in contact with the flesh — from his head to his tail, so as to separate the blubber. You have an even quantity on each side. If the elephant is fat and has nice thick blubber, you commence at his head or tail and cut pieces large enough for a man to carry in one hand. As you cut it off you run your knife through it and it serves for grasping a hold of. After having taken all of the blubber off the back you turn him over on his side and go through the same performance. It takes eight or ten men to turn a good sized bull over. The men stand by with poles, two men to a pole. The pole is shoved through the holes already made in the blubber; two men can generally carry six or seven pieces. It is conveyed to the beach and from there rafted[9] aboard to go through the process of trying, [or] boiling out. Skinning and conveying the blubber to the beach is kept up all day.

About 4 o'clock P.M. we had got about all the elephants killed that were to be seen; only here and there could be seen a stray one. Having got all of the blubber down to the beach, we proceeded to take it aboard in the boats. We worked hard until nightfall, and still there was a large quantity to bring. We went aboard that night and after a good hearty meal we turned in to sleep our weariness off. Next morning found us at it again, bringing blubber aboard. Part of the men were left aboard to mince the blubber, that is to cut the chunks down each side the opposite way so as when it is tryed out, the heat will have instant effect upon it. It had been decided not to try it out at the present, but after mincing it to put it into casks to [await] further orders. The men were kept busy all of the morning and part of the evening. At last we had got as much as we could possibly accommodate aboard, but as yet the beach had not been cleared of it. We were sent ashore to bury it in the sand. The reason for doing this is because the birds will fly down and devour it.

The most voracious bird, I believe that I have ever seen, was on this island, the nellie, or better known as the stinker or turkey buzzard of the Antarctics, and also the sea hen is very bold, in fact, I believe, the most avaricious of the two.[10] If it happens to see you approach its nest, it will pounce on you, and often inflicts serious wounds with their powerful beaks. It's about the size of a domesticated chicken. The nellie is also a very bold bird, but not given to warfare. About half a dozen of them will devour an eight barrel elephant in a very short space of time. They will stand around the carcass of an elephant, gorge themselves until they can't eat any more. You will see them vomit it up, take a good wash, and fly back and get another feed. They will repeat this until every speck of blubber is wiped out. I noticed one thing curious: that is, when we commenced to kill the elephants, there was not a nellie to be seen. We had not finished killing our first elephant before they commenced to swarm around us. The smell of the gore must have been wafted by the wind to their haunts. The nellie attains the size of one of our large turkeys. Its color is jet black. The sea hen is of a chocolate color. Both of these birds are very swift of flight. I noticed the stinkers flying. [They] are found a thousand miles from land sometimes and they skim through the air at such a terrific rate that you can hear their wings make a peculiar sound, loud enough to be heard a good ways off.

By 3 o'clock P.M. we had all of the blubber buried, so we all went aboard. We were to head that evening for Table Bay, in quest of more elephants. After arranging things aboard, anchor was hove up, and we set sail. The distance was some twelve miles to the eastward. We were until sunset making the harbor. That night as usual when in harbor, an anchor watch was set and all hands turned in.

Next morning early all hands were ready to commence work again in earnest. Both boats were launched, and we pulled for the beach, had no difficulty in landing. After every thing was landed, we made our way to the farm yard, as it was called, on account of it being such a nice locality for elephants and penguins. We had not proceeded down the beach far — a very broad and level one — when a grand sight presented itself to view. Multitudes and multitudes of penguins could be seen strutting about in regular military style. A person would imagine that they were trained to tactics, regular files and lines of them marching up and down, sentinels standing as straight and upright as soldiers. This gave them the appearance of soldiers. This was a king penguin rookery, the largest species of the class of birds. They range in height from two to two and a half feet high. They are a magnificent bird; their plumage is very beautiful and bright. Their feathers on the back are of a dark blue velvety hue. Their breast is of an alabaster white, their head is of the same color as what their back is. They have a regular formed line running from their bill down across their shoulders. It gradually disappears towards the bottom of their body. The legs are as black as ebony.

I and a couple of my mates walked up to where they were all congregated. They did not seem to be very timid, in fact they let you walk amongst them. One of the men got a hold of one of them, put his hand underneath its tail and pulled out an egg; this seemed to be quite a novelty to me. We killed a few of them. I evinced a good deal of interest about the curious way in which they have of carrying their eggs. After subjecting one to an examination, I found that they are fitted out with a bag, right under its tail. They carry their egg — they never lay but one — [in] this bag.[11] The flesh of the young is very palatable, and their eggs are as good as those of any fowl. When you cook the eggs, the albumen in them does not turn white like that of a fowl, but it has a translucent appearance. They are splendid eating.

After satisfying our curiosity we went on down the beach to where the rest of the crew had gone. Arriving there we found that they had commenced to kill elephants. They had a considerable number killed already as at this place they seemed to be very plentiful. After going through the same performance as what we did at Sprightly Bay (that is, having disposed of our blubber, it being the last of the elephant season), there was no more to be found. We ran down to Mutton Cove,[12] laid in there three days and from there we went to Three Island Harbor on the S.E. side of the island. We were not very long in wending our way there.

Upon arriving there we found quite a number of ships lying at anchor. Three vessels had come over from Heard Island, four barks and five schooners. They had fixed upon this place for headquarters and a'laying in supply place. The bark *Dove*[13] had discharged a cargo of supplies there, and these different vessels were getting in a stock of stores. These vessels were out on the same mission as what we were, getting oil. It was quite a treat for us to meet with so many vessels. After getting some provisions, we departed and made our way to Pot Harbor. Arriving at this place we found the bark *Dove* and the schooner *Exile* of New London.[14] Our reasons for going to this place were to see about some provisions. We let go anchor there and laid over for a week. During that time we did not do much to speak of. We were aboard of the *Dove* and *Exile* and got some books and papers in exchange for some that we had read. At the end of the week we weighed anchor and headed for Three Island Harbor.

We had received notice previous to leaving Pot Harbor that we were to go right whaling.[15] Having proceeded up as far as Royal Sound, the look out was heard to sing out, "There she blows, two points on the weather bow." Everything was excitement then; the boats were lowered. After a good deal of rowing about he was captured by the other boat's crew. He was a good sized fellow netting us about forty barrels of oil and six hundred pounds of bone.

After cutting him up and taking his bone and blubber aboard, we made way for Three Island Harbor.

We selected a place of anchoring, where the wind would not strike us too heavy, as we were going to commence to try out the blubber we had aboard. The anchor was let go accordingly, and we commenced operation. As fast as it was boiled out it was put into casks and taken ashore. We then took it [above] the water mark on the beach and stowed it away. It had to be covered with earth as the sun would have an ill effect upon it if it was exposed. The captain did try to send it home, out of the three barks and three schooners that left. He was not able to get any of them to take it home, as they were all heavy laden.

After disposing of this lot of oil in the manner that I have stated, we commenced to get ready for whaling in earnest. From here we went back to Pot Harbor by the way of Royal Sound. The reasons for doing this, I learned, was to be on the whaling grounds in the right season. We were not there long before we captured three good sized right whales and three humpbacks. We thought that we were going to have a good season of it, but in this we were disappointed. They did not turn out very abundant. Disposed of our blubber by mincing it and putting it away in casks. We left Pot Harbor for Table Bay to go elephanting again. We were to go there as it was thought that they were very plentiful there. Arriving there we found it to be the fact. We managed to get the vessel full. We then went back to Three Island Harbor, to try it out and see if we could not manage to ship part of it home. Upon arriving there we found the bark *Monticello* from New London.[16] She had brought for us some letters. I heard from home, and [to] my great relief that every body was well.

The captain received a letter from the owners to the effect that [they] wanted him [to] stay out a year longer. Some of the foremast hands upon hearing this grew frantic. They said that they had only shipped for one year, and they demanded to be sent home if he was going to carry the agents' wish into effect. The captain went and got the articles and showed it to the men. They came to find out that there was no specified time on the articles.[17]

The men, at least part of them, were rather crestfallen over this.
The captain then commenced to ask the men if they wanted to
stay. He came to me and ask if I would stay? I told him yes providing he would send two men away that were making things very
unpleasant in the fo'castle. I then told him that I was certain if he
could manage to get these men away the rest of the crew would
remain. These two men were the roughest kind of character, always quarreling or fighting. In fact they made things very unpleasant to everybody, as the rest of us believed in living in harmony. The captain seen that the best course he could pursue was
to get rid of the men, so he came and told us that he would consent
to the agreement. The two were accordingly sent away on the
Monticello bound for New London and we lived in harmony ever
afterwards.

During the next two days we were busy getting the vessel
ready for whaling. We, at length, got things fixed up, and we
weighed anchor and sailed for Pot Harbor again. Cruised around
for three months, meeting with various successes. We then went
to the Island of Rodrique.[18] It lies in Latitude 19°40' South and
Longitude 63°24' East. The island is rather small, about twenty
miles in circumference. It is very high. There is a reef that extends
very near around it. When you come to an anchor you are obliged
to run right between two reefs, until you arrive inside. The island
looked very green and beautiful to us after being in cold weather
so long. The island was originally settled by Rodrique, the noted
pirate. It is now an English possession. The inhabitants are
mostly French creoles, a very indolent and not at all progressive
class of people. It is very fertile and productive; in fact everything
seems to grow spontaneously, all kind of fruits and vegetables can
be procured at this place. The captain gave us all a run ashore, and
we enjoyed our self very much.

After a stay of a week we left to cruise around the island in
quest of sperm whales, but unfortunately we got nothing. We
were just thinking about leaving the place when we fell in with the
bark *Milwood* of New Bedford.[19] She was found on a cruise to
Madagascar looking for humpback and sperm whales. We accompanied the *Milwood* all the way. We were on the "look out" for

whales but did not see anything. We arrived at St. Mary's, a small island about ten miles off the coast of Madagascar and came to an anchor in the harbor. This place is not of much importance. It is adorned with a fort and government house. There is also a dock yard, and at the entrance of the bay there is a small island; the governor and commander in chief of army reside on it. Opposite to this island on the mainland is a French settlement of some ten houses. They have a church. They were trying the experiment of making sugar. After staying there a couple of weeks we weighed anchor and went around to Antongil Bay. It was [thought] that we would encounter plenty of humpbacks in the straits that flowed between the two islands. We run into Port Choiseul and anchored. Three days had elapsed and we had not come across any humpbacks. We then decided to cruise about. On the fourth day the bark *Milwood* parted our company, after having got disgusted with the whaling grounds. On the third day after her departure we captured a humpback, and we seen plenty of them, but our boatsteerer did not understand the business. We did not get any more. The men foreward prevailed upon the captain to put some one in his place, but the mate put it into the captain's head not to, consequently we lost a good many whale. He finally did so, but it was too late in the season.

A few weeks afterwards we returned to St. Mary's and the men got a run ashore. We took in a supply of wood and water, and we were to give the vessel a good [cleaning] up. From here we were to return to Desolation Island. We received the news of the rebellion at home [the Civil War]; I assure you that it made us feel very anxious about our people and homes. All of us were very eager to get back home. But no, we had to go to Desolation.

On the 3rd inst. we set sail for Desolation. We had fine weather and all went well until we got within three days of [the] island. We encountered the most terrific gale that we had in the whole voyage. It lasted for about two weeks, a terrible sea running all the time. The gale at last partly subdued, and we managed to get to Desolation. We anchored in Sprightly Bay. We managed to fill the vessel up in a short while. We then left this place bound for home by the way of Cape Town. Enjoyed fine weather the whole passage. Arriving at Cape Town, we let go anchor in Table Bay. We went ashore. It is quite a place, and I might say the only real decent place we had seen since leaving the States. I was very favorably impressed with the place. It is quite a commercial center, enormous quantities of ivory, diamonds, gold, and ostrich feathers are shipped from the place. At the end of a week, we weighed anchor and sailed for Home Sweet Home. One of the men was taken very ill, so we run into the Island of Saint Helena; we let him off there and as the vessel was leaking so much we shipped two hands to help pump. After a pleasant passage of forty-six days we arrived home in New London.

When I come to settle up accounts with the ship owners I found myself $42.00 in debt to them. I was rather disgusted at this. I must say they then offered to pay my fare home, but I declined the offer, and I defrayed my own expenses there. My people were very glad to see me home again, and wanted me to remain, but after a stay of three weeks, I made up my mind to go and join the navy, and fight for the Union. Men were in great demand in the service and it was a time when every one was called upon to fight for his right. Having my mind made up to this effect, I informed my father and mother about it. They seen that it was useless to try and persuade me not to, so I was let go without further delay.

Four days later found me in Boston, Mass. I went to the [illegible] and enlisted for a space of two years on the guard ship of Charlestown. From there I was drafted to the gunboat *Genesee*. We went from Charlestown to the James River. After a stay there of three months, we were ordered to Wilmington, N.C. to help on the blockade. We stopped there for two months, and then we were ordered to the Mississippi River. We remained in the River until all of the Confederate forts were taken possession by the Federal forces. We were then ordered to Ship Island, a small island lying in the Mississippi Sound. We were in the blockade and did duty until it finished. My time had now expired. I received my discharge and after bidding farewell to my ship mates, I was taken aboard of the transport and taken to Philadelphia. I once more made for home; two weeks later found me at home.

Had been home but a short while there when I received a letter from Williams Haven & Co., asking me to join a schooner that
was fitting out for Desolation Island, the *Roswell King*. I was to go
as boatsteerer, read the letter, if I agreed to go to come down to
New London immediately. As everything was dull around home,
I thought I would try it again. Two days later I was in New London. The first two personages I met were Mr. Chaplin[20] of the
firm and the captain. Of course they wanted me to go down to the
office and sign. I had not fully made up my mind to ship, so I told
them that I'd be around to see them next day. So next day I went
down to the office, fixed matters and shipped as boatsteerer
aboard of the *Roswell King* bound for Desolation Island. As you
see, I had taken one step toward the height of my ambition. I had
now made up my mind fully to become the master of a vessel. My
future career was to be worked in this view.

I will not enter into details regarding this voyage or the succeeding one. Suffice to say that they proved successful financially,
and that on the succeeding one I realized the height of my
ambition — for the time being — I became first mate. I shipped
for a space of three years. We went on a cruise to Desolation Island. In the third year Captain Glass was taken ill. It seems that
he had been ailing from some chronic complaint before. He wrote
home to inform them that if they did not send some one out to
take his place, he would go home. Some months afterwards the
freighting vessel arrived, and it bore Capt. James Church, who
was to take command. I was tempted to return in her myself, for
according to the ship's articles they could not oblige me to remain.
By right the ship should have been sent home. I asked the different captains' advice and they told me that it would be to my interest to remain. I did and I assure you that I never regretted it. If I
had done as I was tempted to do at first, my bright future would
have passed on to oblivion. But thanks to fate I did not.

Captain Church accordingly took command of the vessel, and
I assure you that I do not say too much when I say that Captain
Church was acquainted with every nook and corner about the island. My cruise with him proved very beneficial to me in after
years, as he made me acquainted with parts of the island that I had

no knowledge of. The voyage proved a success. We got a good cargo of elephant oil. We returned home. Captain Church spoke very highly of me, so the owners offered to fit the vessel out again and give me the mastership. I accepted it, of course. Now I had reached the utmost height of my ambition. Scarcely twelve years had elapsed since I had commenced to go to sea and now I was the master of a vessel. By constant attendance to my duties and by applying myself to the study of navigation, I had accomplished what I then realized.

All arrangements were made for sailing in a short time. We were to go to Desolation Island in quest of oil, either whale or elephant. During my stay at home I got married to a Miss Jane M. Adams of New London. I sailed shortly afterwards to be gone three years. To tell the truth I did not stay at home to enjoy my honey moon but almost directly after taking to myself a better half, I departed for Desolation. I did not return for three years, so you can see that I did not enjoy much connubial happiness during [the] first part of my married life.

The voyage was a successful one. I got about two thousand barrels of oil and the same weight of whalebone. I left the island about a month before the expiration of the three years. I went to Cape Town, South Africa. I there found my mother-in-law. It seems that her son had prevailed upon her to come there, and she did so, but getting disgusted with the country she was eager to return to the States. So she accompanied us home. Arrived home after a fair passage. The owners were very well pleased with me, and I myself was highly elated over my success. I must say, they offered the vessel again to me. This time I was to engage in another branch of the business. I was to go by the way of the Crozets,[21] and see if I could get any seal, as the prices for seal skins were away up. And it was thought that they were very abundant on these islands.

In due time the vessel was fitted out for sea. I sailed from New London with 16 hands, all told. We were to get to ship eight more hands upon arriving at the Cape de Verde Islands. We also had a passenger for the Island of Bravo. This island is the most south westerly of the group. It has no harbors or anchorage. The beach is a very rugged one. The island seems to be very fertile. We laid off and on about the island. Of course, I had to go ashore and see about getting my men. The town does not amount to much. Its most numerous inhabitants are blacks intermixed with the Portuguese. The blacks are originally from the west coast of Africa. The island derives its name from Cape de Verde on the African coast. Having got an addition of eight men to my ship's crew, I sailed for the South Seas. After a voyage of fifty-five days not encountering much foul weather, I arrived at Hog Island. This island belongs to the Crozet group, there being five in all, and it lies 46°2′ South, 50°10′ East. I anchored here, as I was going to try my luck at sealing.

As I wanted to see the island myself, on the afternoon of the second day of our arrival there I had the men lower the boat, and accompanied by twelve of them we went ashore. We experienced much trouble in getting ashore, as the beach is very rugged and uneven. At length we managed to land. I sent part of the men to reconnoitre about the place, but they came back saying that they could not see any seal. I had them launch the boat again, as I had had it pulled up on to the beach, and we then went back to the vessel. We had no sooner arrived aboard, and it commenced to blow heavy from the N.W. It blew for three days, and on the third day it subdued.

While at breakfast next morning, the first mate, Mr. Usher, and the second mate, Mr. Joseph, asked me if I would not allow them to take a boat and go ashore. I told them that if they thought that they could land they could take the boat. They said that they were confident that they could land. So after cautioning them in regard to the sea running around the island, and I knew myself that it was very rough, everything was got ready and they pulled ashore toward the island. I thought no more of it for the time and went back down into the cabin after telling them what to do before leaving. A half an hour had scarcely fled before one of the foremast hands came running aft to tell me that the boat had capsized on the beach.

I immediately went on deck with my glasses and took a view of the place. I could see the men hauling the boat on to the beach, so

I thought that all was O.K. with them and as they did not
signalize for help, I dismissed the thought that any harm had
come to them. It got along towards five o'clock in the afternoon
and as yet they were not putting in an appearance. I thought that I
would not wait any longer but I decided to go and see what the
matter could be with them. I had no intentions of landing as I
could see it was very rough on the beach. I informed Mr. Glass,
the third mate, of my intention, and then I took boat's crew and
pulled over towards the island. We pulled dead for the island. I
could see tremendous big breakers running in and the surf for a
ways around was white with foam. I was going to try and get up
as near to the island as possible, that is, within speaking distance.
It was a perilous undertaking, but it had to be done. The casta-
ways seen me coming. I had the men to pull up as close to the
beach as possible. It was a wild and merciless looking one.

At last I got within speaking distance. I hailed the mate, and
asked him what had happened. He replied by telling me that he
had stove in the boat, and that he had lost four of the boat's crew.
I felt rather bad about this, I must say. I then told him that I
would not risk landing that night, and that he would have to make
the best of it until morning and that if everything proved favorable
I would land in the morning and take them off. After doing this I
returned to the vessel. I had the crew all muster aft, and I found
that he had twelve men ashore with him. I then dismissed, telling
my boat's crew that we would go ashore in the morning. I must
say that the loss of the four men weighed hardly on my mind.

After breakfast next morning I had the boat lowered and
pulled off toward the island. Before doing so I gave the third mate
instructions of how to act during my absence. I went toward the
beach where the mate had landed, but upon examining it I found
that it was an utter impossibility to land there without putting our
life in extra peril, so I decided to go down the coast a'ways, and
see if I could see a better prosepect for landing. At last we came
opposite a small beach. It was not so ragged, but still very rough.
The sea did not break as heavy over it as it did up to the other
place. So after instructing the men how to act — as they were all
green hands but one — and what they must do when the boat

struck the beach, keeping a good look out on the sea, and waiting for a favorable one, we tumbled about in the surf like a nut shell. At last a sea favored us and away we went in with it. On the third sea we struck the beach, safe. The men sprang out and grasped a hold of the boat to keep it from going out with the under tow. So the next sea that came along sent us high and dry on the beach.

Having landed, of course, Mr. Usher's party was there to receive me. I gave them some food that I had with me. This they were very glad to get as they were almost famished. We got the boat ready again and tried to get off. We tried three times but were driven back each time by the cruel and merciless breakers. Getting disgusted, I abandoned the notion of trying to get off and took a walk up towards where Mr. Usher had landed on a tour of inspection. Having got up to the beach, I commenced to ask the men how the accident happened. The explanation was not very definite. I then asked Mr. Usher. He laid the fault to the mismanagement of the second mate, who retaliated by laying the fault on to Mr. Usher. I told them that I thought it was mismanagement, and that if things had been conducted right the four men would not have perished, for when the boat got capsized the water was up to a man's waist, and there was a very strong current running out. I myself think that they thought more of the boat than they did of the lives of the men who were lost. Some of them were excellent swimmers and by a little help would have been saved.

After examining the boat, I found that its side was badly stove in. I tried my utmost to get her outside of the breakers, but as soon as we attempted it the breakers would heave her ashore again. We at last gave it up for a bad job, and left her to her own fate on the beach. We then returned to our boat and found the sea running very high. Being very anxious to get off, as I did not know at what minute the wind would chop around to the N.E. and the vessel was in a very bad place, we attempted it again, but with no better success. The sun was just setting. We had a cold nasty night before us, as there was no shelter or no place to rest our weary bones. I thought we were doomed to sleep on the beach that night. Some of the men that had been out exploring the island arrived and told us that they had discovered a house. I asked them the distance to it and they said it was about a mile from where we were. This was very pleasant news to us. I had them haul the boat up and make it fast to some rocks. Then we made our way towards the house. On our way we got all of the young albatross that we could get, as that was about the only food to be had, and they were to comprise our supper, dinner, and breakfast. We managed to get enough birds to do all hands for their meals.

Arriving at the house, we found it in good order and in fact it was a snug and comfortable abode. It was fitted out with a stove and fireplace, one of the regular old time hearths. It being very cold and as [we] were drenched to the skin, we built fires in both of them, fuel being very plentiful. There were a few cooking utensils, such as pots and pans, scattered about here and there. We selected one of the largest pots, gave it a good cleaning out, and then we took and prepared the albatross for cooking. We first skinned them and then we cut all of the fat off of them. This we put into the pot and after the oil had been fried out of it, we took and added the flesh. After letting it cool enough, we took it out. Now all that was lacking was salt. But as the old saying runs, "Beggars should not be choosers." We had to be contented with our lot. They tasted very much like seal. We had no bread but we made up for that by devouring more albatross.

After finishing our frugal but, I dare say, substantial repast, we commenced to divest ourselves of our clothing, so as we could hang them near the fire to dry. Now the next question that arose was, where [were] we to sleep for the night? The house was filled with berths; there was double berths on each side. I at last selected one near the fire in the adjoining room. It happened that there was a hole in it, and I could get a view of what was going on in the next room. So after warning the men in regard to looking out that they did not set fire to anything, I retired. I did not go to sleep right off but I laid watching the men. They seemed to take every thing philosophically, that is, they were resigned to their fate. They were telling stories, and now and then one of them would favour the party with a song; both comical and sentimental ones were sung. At last one of the homesick boys come out with "Where is my wandering boy to night." This made me think of home and my people. It brought back pleasant memories of the past. With beautiful thoughts travelling through my brain, I dropped off into a deep slumber. I threw myself in to the arms of Morpheus.

During my sleep I was troubled with all kinds of uncanny dreams. At one time, I thought I was in the surf, struggling for dear life, and [another] time I would be aboard of the vessel trying to beat away from the island; the cause of this was the large quantity of albatross. I dreamt on in my dream that we were being fast driven ashore. The vessel was just going to hit the beach. Everything was excitement; the men were panic-stricken. At last came the crashing thud, and I awoke. The perspiration was dripping from my forehead. I looked around, dazed and about half-scared to death, at last it flashed upon me where I was.

Looking through the hole at the foot of my berth, I was horror stricken with what presented itself to me: the whole interior portion of the room was one mass of flames. I sprang from my bunk, and at the same time yelling out "fire" at the height of my voice. I dashed through the flames like a mad man. After having collided with numerous pots and pans was very near cutting my head off by running afoul of a line that one of the men had his clothes on, I arrived outside. The other men heard my calls of "fire" and they followed suit by evacuating the premises.

I counted men and found that there was one missing. At last we heard him crying out for help. It seems that he had lost his bearings, and was unable to find his way. Seeing that he would perish if he stayed in there much longer, as the whole building by this time was almost enveloped in flames, I told one of the men to go in after him. He done so; after being in but a short while he came out leading the other fellow by the hand. He was more dead than alive. The other poor fellow was minus part of his whiskers and hair. After feeling around to see if we had any bones broken and feeling of our proboscises to see if we had not rubbed them off against something, we commenced to take in our situation.

Standing there in the glare of the burning house, the light shone on us; it gave us the appearance of beings from the other world, not from heaven but from the infernal regions, we looked like a group of evil spectres. Half-naked we stood shivering in the

cold midnight air. I assure [you] that a Broadway clothier would
have reaped a harvest out of us. We were in a sad plight indeed,
and we had a good part of the night staring us in the face. I asked
the men why they did not take the precaution and why they did
[not] put a man on the look out? I could not get any satisfaction
out of them. Like the first and second mate about the boat, they
blamed each other. My idea of it was that the fire must have origi-
nated in the fireplace. The roof was composed of tarred canvas, so
very inflammable, and to increase the destruction of the house,
one of the men in his hurry to get out overset the stove. It was a
good severe lesson to them to be more careful in the future when
sleeping in strange houses with a couple of big fires a'burning.
The building by this time had been completely consumed and was
now smoldering down into embers. It was a great pity that this
should have happened as the building was of great benefit to vessels cruising in those parts. I found out that it was erected by the
Cape of Good [Hope] Co. some six years previous.

    After watching the building until it had fairly burnt up, I told
the men that we had to make the best of our very unpleasant situation, and that we had better look for a place for shelter. We separated. I and my party picked up some old barrel staves that were
lying strewn about and made our way to where one of the men
said there was a cave. We had walked but a short distance when
we arrived to where it was. Looking in and by the dim light of the
new moon, I could see that it was unoccupied by man or beast.
We entered it and found it quite snug and dry. It was of natural
formation and quite commodious. We made a bed by placing the
staves on the ground. Of course, they were not as soft as down but
they had to do. One of the men said that he believed the feathers
were frozen to the boards. I did not feel at all like laughing at this
bit of humor, as I had commenced to think of the misfortunes of
the day. Being very much fatigued by the happening of the day, I
fell off into a deep sleep.

    I awoke next morning, feeling very bad, and my bones racking
with pain. I awoke the others; they got up feeling about as bad as I
did. We then left our cold and all together wild looking abode. We
went down to the scene of the conflagration; we found that the

rest of the crew were waiting for us. I asked them how they had
fared during the night. They told me that they got to leeward of
some projecting cliffs, and spent a miserable night of it. We were a
sorrowful looking lot of men. All of us now suffered severely, for
the loss of our clothing and now the pains of hunger were com-
mencing to put in an appearance. Dawn had just finished breaking
and we wended our way towards the boat. It was good sun up,
when we arrived there; the boat was all right. The sea had subdued considerable, although it was still very rough. I could see
that the vessel was all right, and it was plain that the third mate
had give her extra mooring chain through the night.

I now made up my mind to get off, come what may. I asked
the men's opinion and they seemed to be every bit as determined
as myself to get off. I picked out the strongest and most experienced hands who were to man the oars. The others were to get
into the boat after she had been floated, and get out of the way, by
stowing themselves under the thwarts, and in case of necessity to
bail out. I had them all arranged, each one in his assigned place. I
divided them by having an equal number on each side, so as to
steady her in the water. I had them to haul the boat down to the
water's edge; we were up to our waist in it. The men were now in
their assigned places, and I had to give them orders what to do. I
had the oar locks put in their place and had the oars shipped so as
the men could grasp a hold of them immediately upon springing
into the boat. As a general rule in surfing, this is always done as
the men would be incommoded upon getting in.

Everything was now in readyance. I assigned the two
boatsteerers to the forward bow oars, Mr. Joseph, second mate,
the amidship oar and two of the best men to man the other two
oars. Generally there is three heavy breakers and then two smooth
ones between. I gave the bowmen orders to get in first and take a
hold at their oars, and then the rest of us followed suit. Mr. Usher
was to be the last one to get in. He was to jump in the stern as the
boat went out.

I gave them the order for the boat to be pushed and we all
scrambled in and went out on the second smooth sea. Could see a
third one a'coming and a monstrous big fellow he was. We got

over the first and second ones all right. She rode them like a duck.
But the third one stood her right up on her head and broke right
into us, filled up the boat half full of water. The men that was underneath did not seem to mind it as it seemed a morning ablution
to them. I sang out to the men to lay back and they pulled away
for dear life, knowing full well the necessity of it. Gradually we
got from three or four ship's lengths from the cruel and merciless-
looking beach. The men's muscles then relaxed and their excitement grew less. I told them to take a blow; they had exhausted
themselves to such an extent that they were pretty tired. The
majority of us had stripped ourselves to the skin so as to be ready
in case anything happened. Our excitement had been thrown up
to such a pitch that we never felt the effects of the cold chilly air or
icy water.

After having a good blow we resumed our journey. After a
good deal of hard pulling we managed to get to the vessel. We all
got aboard safe. The third mate seemed very glad to see us. I
could see from his appearance that he had worried himself a good
deal about us. In fact he had not slept a wink through the night. I
can assure you that we all offered up a prayer to the Almighty for
our kind deliverance. Of course we were almost famished to
death. We were supplied with a sumptuous meal from the ship's
larder. We could appreciate a good meal, after having had to eat
albatross without salt or bread. I tell you that a slice of "the staff of
life" was quite a luxury.

We were fortunate in getting aboard at the time we did, as we
had been there but two or three hours, when it came on to blow
very heavy from the N.E. The wind seemed to be steady and once
in a while it would lull down so we were not afraid of her dragging
anchor. It kept blowing about the same the rest of the day and the
[succeeding day] it kept up about the same. At dark it commenced
to come on heavier. The vessel was pitching and heaving about;
large volumes of water were constantly coming over the ship's
side. The night was a dark and rainy one, the surf was increasing
in size, and the whole ship was drenched in water. The wind
commenced to increase. I could see that it was fool-hardiness for
us to remain overnight, as there was no prospect of the wind fall-

ing, and I did not know at what moment she would drag anchor. And it was useless for us to try and heave it up. I gave orders for the men to put double reefs in the foresail and main sail and to take the "bonnet" out of the jib. I then sent the first mate forward, with instructions to knock the pin out of the shackle when her head paid off shore, and also to have a buoy attached to [the] chain ([we] used [a] barrel).

The wind had now increased in strength, and it was blowing next to a living gale. At length her head commenced to pay off shore. I gave the order for the anchor chain to be let go. It was obeyed in an instant. As soon as this was done, I had the jib cracked on. I had the foresail and mainsail double reefed put on; we then went out on the wind. I then stood away from the land. It was cruel work, but indeed our only salvation. Our reason for letting go our chain was because it would have been an utter impossibility for us to have hove it in, as the enormous strain that would be brought to bear on the windlass would tear it out of its position, hence our reasons for slipping anchor.

We at length managed to get clear of the island, but we met with other crosses. The wind now was blowing a living gale, and an enormous big sea was running. And to make us feel more miserable, it came on to rain in torrents. It was useless for us to try and make any head way that night so I had jib and mainsail furled. I then had them put on the storm try sail and we laid to for the night. Everything was all right until morning. We headed east south east all through the night.

Next morning the wind hauled to west north west. It was still blowing a strong gale. I kept her off and headed for Desolation, which lies about one thousand miles to the E.S. eastern of the Crozets. It took us five days to make Desolation. We anchored in Sprightly Bay. Once more back to our old haunts. The old and mournful cries of the penguin was like music to my ear as it brought back old memories of the past. I noticed while passing Cape Bourbon the multitudes and multitudes of them that could be seen on that place.

The rookery is about two and a half miles in circumference, so you can fancy the enormous quantity there must have been of them, uttering their mournful notes, marching about like soldiers. All in all it was a grand sight to view. These are unlike their brothers, the king and johnnie, in appearance. They have a tuft of feathers of bright yellow hue on their heads, giving them in connection with the way they have of marching a military appearance. The color of their body is identically like that of the king.

There were no other ships in Sprightly Bay but ourselves, so we were lords of all we surveyed. We were to go elephanting next. We went ashore and met with success, as also we did during the preceding times. Remained there one month and got four hundred barrels of oil. This we conveyed to the Three Islands Harbor. We found the bark *Roman*²² there and I succeeded in sending it home on her. I also got a chain and anchor from them.

The succeeding eight months were spent in cruising after whale in Pot Harbor and Whale Bay. We did not do extra well as we only got four right and four humpback whales. I decided to go sealing then, so I made for Swains Islands. Arrived there, and after a good deal of trouble we managed to get lightly five resting seal. Finding that they were not at all as abundant as I expected to find them, I quitted the place and made for Terror Reef. We were about five or six miles off of Terror Reef when we sighted a steam ship. She was taking soundings. She signalized that she wanted to speak to us. I then kept off and run down to her.

Upon nearing the vessel, I got a glimpse of her hailings and found that she was an English ship H.M.S. *Challenger.*²³ We run around her stern and came up alongside her larboard side. The captain was on the poop deck with his officers. He hailed me and wanted to know [if] that was the *Roswell King*, Captain Fuller? I assented, saying yes. He then asked me if there was a harbor where I was going to anchor for the night. I told him that I had not decided upon any place as yet.

"Well, captain," he said, "with your permission I will go aboard as I would like to speak to you on a matter of importance."

I told him that I would be only too glad to be honored by a visit from him. He had his boat lowered, and he came aboard. I was in the waist to receive him. At first he did not pay any attention to me. He seemed to be looking around for somebody else. "Captain Fuller?" he said.

"Yes sir," replied I. "I am he."

"Ah!" exclaimed he, "you will excuse me." On account of my youthful appearance, he did not at first think that I was the master of the vessel. "You will excuse me, as I expected to see some old man, but I have before me a young man just in the bloom of manhood."

He grasped me by the hand and gave it a hearty shake. I felt rather embarrassed and to use a "Jack Tar" phrase, "I had the wind taken out of my sails." He asked me at what age I had commenced to go to sea. I told him.

"Well," he said, "you have made broad strides. You must have been a young man of ambition."

I invited them down into the cabin, the captain and his lieutenant.[24] They accepted the invitation. I had made up my mind to entertain them the best that I could. After getting them seated, I told them that they must not mind the very unelaborate appearance of the cabin. It was unlike the kind that they were accustomed to set in. And that my wines were not of the finest assortment. In fact the only liquor aboard was some Kentucky whiskey, and they would honor me by accepting a glass of it. I had the steward bring some. They filled their glasses and Captain Nares arose and offered a toast to me, that my voyage would be a financial success. I retaliated by offering a toast to their success in the voyage that they were undertaking. I proposed the toast with Adam's ale, as I was strictly temperate.

Having finished toasting each other the captain informed me of his mission to those parts. First of all, it was a voyage in commemoration of the voyage of Captain Cook. He was also looking for a locality to observe the transit of Venus.[25] In all, he was out for the benefit of science to all nations.

He wanted me to give him some information in regard to the island.

"You have a chart of the island, I suppose, captain," said Captain Nares.

"Not exactly," replied I, "that is, I have not one drafted but I have one in my head."

I told him that I was willing to impart all of the information he wished me to. He then wanted me to show him some harbor where he could anchor for the night. I then told him that there was Little's Harbor about 13 miles from there, and that he could get good anchorage there. I myself was going to run in there for the night.

"All right," he said, "I will go in there myself."

After saying a few more words in regard to different things, and asking me about the bearing of different positions of the island, he and the lieutenant started to go up on deck. As he was just going to ascend the companion he remarked [that] like I had said, my cabin was not an elaborate one, but he assured me that it was good and commodious for a craft of its size. We then went up on deck. As he went down in to the waist, he said, "Now, captain, I will expect your company at dinner. I want you to promise me that you will come." I told him that I would go with pleasure. "All right then, captain, and by the way," continued he, "you had better take in all your sails and the *Challenger* will tow you in."

I thanked him very much but declined the offer of being towed in. After getting in his boat, he and his lieutenant waved their hands at me, and he said, "*A* [Au] *revoir*, captain, and don't forget, 6 o'clock P.M." They had just left and I set sail, and kept away for Little's Harbor. I came to an anchor there at about 5 o'clock P.M. and the *Challenger* came in about an hour later. She had no sooner let go her anchor than the boat was sent off after me. I went on board in company of Captain Bailey of the schooner *Emma Jane* of New London.[26] She had been lying in Little's Harbor for some time. Having arrived on board, we were met by the officer of the deck and conducted into the mess room. We were received by the captain and his officers and were set down to a sumptious meal of victuals. During the repast, Captain Nares gave me an account of his passage from the Cape of Good Hope to Desolation. Lieutenant Haines also made himself very agreeable. We did not tarry long at the table, as I had told the captain that I intended to sail next day if everything proved favorable. He told me that he would not detain me on any account, but would proceed to business.

He asked us to stop up into the upper cabin. Having arrived there, he got his charts of the island out. And from seven until twelve o'clock I was kept busy pointing out this point, that reef, etc.; they had no authentic chart of the island. As there were a good many important points to be observed when cruising in and about the island, the captain thanked me very much and complimented me for my knowledge of the island and its surroundings. As he said, the information that I had imparted to him would prove of the utmost importance to him. Before leaving he asked me if there was any article that I was standing in need of and if so he would be glad to recompense by giving me whatever I was in need of. I thanked him very much for his likeable offer, and I declined to take anything.

"There is no use in you trying to get out of it," he said, "I am bound to repay you for the very important information that you have gave me in regard to the island. And you will have to accept some gift from me. Now, how would a half dozen bottles of Jamaica rum suit you? But, I forget, you say that you intend to leave early in the morning. Now as it is past twelve, and as I cannot get into the wine room, I will have it sent aboard of your vessel early tomorrow morning."

I told him that as I would sail before anybody was astir there was no use of him going to the trouble in getting it out and besides I did not care for it. He still would insist upon me accepting it. It had just struck one when I left the *Challenger*. I had just got into the boat, when the captain asked me if there was any place on or about the island that had been named after me. I told him no. "Well," he said, "I will have some point on or about the island named after you and will see that there is a mention made of it on the chart."[27]

I thanked him very much, bid him "good night," and the boat pulled me aboard of my vessel. Next morning at daybreak, I had the anchor catted, it being splendid weather, and a good fair breeze prevailing from the N.W. We was to go a'cruising in quest of right whale. We sailed within speaking distance of the *Chal-*

lenger. The captain was on the quarter deck; he and his officers
waved their hands at me. The captain then told me that he would
give that half-dozen bottles of rum to my friend Captain Bailey of
the *Emma Jane*, and he also said that when he went to the southard
a'cruising if he happened to notice any islands with seal on them
he would notify me. I thanked him very much. He then said
"Good-bye, captain, may you have excellent luck in your undertaking."

I bid him good-bye, and we sailed on out of speaking distance.
I dismissed the occurrence from my mind, and settled down into
my old routine of life. The first mate, Mr. Usher, overheard the
words I and the captain of the *Challenger* had had together, but he
did not seem to put any reliance on them. After cruising around
for nine or ten days, I came to anchor in Exile Harbor.[28] I found
that the *Emma Jane* had arrived previous to me. I was up on the
quarter deck looking at the *Emma Jane*. I and the first mate had
been conversing together, and I was just going to go below, when
he said, "Now you have an opportunity to get that rum that the
man-o'-war captain promised to send you." It had completely fled
from my mind.

"Well," I said, "You see that everything is put straight aboard
and we will go aboard of the *Jane* and have a chat with the captain." A half an hour afterwards, the boat was lowered and we
went aboard of her. We were invited into the cabin. After talking
on different topics for an hour or so — in this time Captain Bailey
had made no illusion [allusion] to the liquor — we were about to
leave, and I let out a passing remark about it. I could see that he
was trying to get rid of us in a sneakish manner; seeing that he
wasn't going to broach the matter, I told him what the captain of
the *Challenger* had told me. At first he did not want to give into it
but he seen that I had him cornered, and he admitted that the captain of the *Challenger* had gave him a half-dozen bottles of rum to
give to me. "And where [are] they?" said I, "you don't think that I
am going to let you keep them."

He then spun me a long yarn about the mate and steward getting into them and drinking them all up. I could see from the ap-

pearance of his face that he had been indulging in some kind of liquor, so I was convinced that he was the culprit.

"Look here, captain," I added, "I consider this a breach of trust on your part. If you had a spark of principle about you, you would have never lowered yourself to this. There is no use of you trying to get clear of it by throwing it on to your mate, as your face and general appearance is conclusive proof that you have been indulging in a debauch. Of all the contemptible, mean, currish tricks that I ever had played on me, this caps them all. The idea, a master of a vessel to be guilty of such a paltry act. You can consider our friendship from this time on at an end."

Saying this, I got into the boat and had them pull me aboard. Bailey stood watching us as we went off. He seemed to have been transformed into a statue. I afterwards found out more of his low character and I came to think less of him. I heard that he owned up to it being a low, contemptible trick afterwards.

After the departure of the *Challenger*, we went a'whaling. The winter's catch proved to be very good. After this we got everything in readyance for going around to the west side of the island. Several [weeks later] we left off whaling and were making preparations to go around to the west side of the island. We were to go through the Marienne Straits into Young William's Bay. I had been there six years previous, and I knew full well that it was about the most dangerous and nasty place about the island. If you are desirious of reaching Young William's Bay, you have to go through the Straits of Marienne, and while going through them if the wind is not favorable — as you have to get in from the north, so as the water will be placid — you have to drop your anchor and lie there until the winds favor you. You then go out on the west side by the way of Cape Louis. These two passages are exceedingly narrow and are very dangerous to be in during a blow, as there is not much sea room. The latter place of the above mentioned ones [is] the worst for there is a very nasty sea running up it continually.

Going through the Straits to William's Bay you pass between a breaker called "Old Bull" and Muscle Bay Point. In fact the whole

region there abouts abounds in reefs and rocks. Upon reaching
William's Bay, you have another unpleasant obstacle to contend
with. There [is] a very large bar at the entrance of this bay and
you have to wait until the water is smooth to fit over it. And if it
happens that the wind is strong out of the bay, you are obliged to
anchor at the mouth of it, and then you have to kedge in with the
use of boat anchors. This we done by taking an anchor in a boat
ahead and casting it. You then bend on a line and it runs to the
vessel and they aboard put it on the windlass and heave, which
brings her ahead. This is repeated until they arrive where they
want to. Having arrived inside you drop anchor in twelve fathoms
of water. You then pay out ninety fathoms of your mooring chain.
Then your vessel will lie in three fathoms. When lying in Young
William's Bay you are obliged to use two anchors down. You cast
one on the bay and the other out of the bay.

The bay is situated between two mountains in a perfect valley.
When it blows down this valley, the wind comes down so heavy
that it lashes and beats the water into foam, causing quite a sea to
come on. The head of the bay is lowlands; consequently the
mountains cause the valley to act as a kind of a funnel. At the terminus of the bay and around the point there is a Bull Beach. It extends for about one mile. It is here that the sea elephant come up
to have their young. I dare say that they come to this place in
thousands. I and my ship's crew have slaughtered as high as 300
bulls and the same number of cows.

Pardon me for having gone off the main strain of the story. I
will now continue. We made away for Three Island [Harbor], arriving there and put our winter oil ashore, so then run into Shoal
Water Bay for the night. Early next morning we got under way
and stood off for Exile Harbor. As we were rounding Cape Digby
the wind came on northeast and blowed pretty heavy. We run
into Exile Harbor and laid there for two days. On the second day
we got up our anchor and were intending to go around to the west
side, but it came on to blow and we took refuge in Fuller's or Little's Harbor. The next day turned out to be a fine one. So we got
underway and worked up to Carroll Cove or Breakwater Bay,
where we landed thirty barrels of blubber that we had got on com-

ing down. We were expecting to have a continual spell of fine weather but in this we were chagrined as it came on to blow very heavy from the northeast. Consequently we were obliged to stay there. The men wanted to go ashore and procure some coal from a "run," a mine. As I had told them that we were going to experience bad weather on the other side of the island, our fuel being scarce, they wanted to get some coal for the fo'castle fire. I gave them permission to go ashore. They returned shortly afterwards with a good supply of fair quality coal.

Next day turned out to be a favorable one. I had the anchor taken up, and we stood away for the west side of [the] island, but it come to blow from the northard and we had the wind dead in our teeth, so we run into Christmas Harbor. This place is the most noteworthy of interest on the island as it was here that Kerguelen, the discoverer of this island, anchored in fact on a Christmas day in the year of 1772, and Captain Cook [succeeded] him four years later.

The ill-fated Kerguelen afterwards committed suicide by blowing his brains out. This is the way the story runs. It seems that when Kerguelen came into the Christmas Bay Harbor, he was under the impression that he had discovered the southern continent. He left Christmas Harbor and headed S.S.E. After getting around the promontory that is known as Cape Digby, [he] headed S.S.W. and [continued] until he came in sight of Kerguelen's Head. He was now convinced that he had discovered the southern continent. He then headed for the Antipodes. Returning to France, he repeated his discovery to the French government. They asked him if he had made any attempt to circumnavigate the island. He then told them that he had gone down as far as the foreland. He then could see land on down to the southern westard and thought that it was not necessary to go any further. The government was not satisfied with this, and they ordered him to return and get some more authentic account of the island. He accordingly returned and started on a cruise around it. He rounded Kerguelen Head and came in view of the extreme land known as Cape Bourbon. He felt downhearted at this, as he had high hopes of doing some great deed for his country, and as to return to

France and be ridiculed he would not and maybe suffer imprisonment. He was driven to desperation; he went into his cabin, took a loaded pistol, and put an end to his eventful career. So ended the life of one of France's greatest navigators.[29]

We let loose our anchor in Christmas Harbor. The wind had not abated as yet and we were obliged to remain in there. On the fifth day the wind lulled down and chopped around to the N.W. Proving to be a favorable wind, we got under way and headed for Cloudy Island. We arrived there in two days. We were just rounding Cloudy Island, when the wind hauled around to the northard, light airs prevailing. We got by the Cloudy Islands by dark, [but] the moon shone bright; we kept her up until we arrived at Cape Louis. Here we let go anchor and remained for the rest of the night. Early next morning at dawn it commenced to breeze up from the S.W. So as soon as daylight came we kept away for Marienne Straits. Arrived at the straits without much delay and let go our anchor.

It proved very fortunate that we got into the straits at the time we did for if we had not we would have been blown away, for scarcely an hour had elapsed before it came on to blow heavy from the west south west. We being in the west passage there was no sea to speak of. The wind had now increased into a living gale. It blew so for almost ten days. Toward the latter part of the tenth day, the wind commenced to fall and by nightfall it was nothing more than a nice fresh breeze. It kept so until next morning. We got under way from the straits and headed for Young William's Bay. By the time we got as far as the passage the wind became very light and as the current favored us, we were not long in getting to Young William's Bay. Arriving at the bay, I found that there was a nice breeze blowing which proved to be fair to the anchorage ground. Having let go our anchors, that is, with one anchor up the bay with ninety fathoms of chain and one anchor out of the bay with sixty fathoms, having got our moorings fixed, I was ready for almost any kind of weather!

We had been there just a week, when some of the men asked me if I would permit them to go ashore and procure some penguin eggs, they being very plentiful about there. I told them yes, but they must not go too far as I wanted them to be aboard before nightfall. They promised faithfully that they would. So I let them go. There was eight of them. I also cautioned them that if they were to go too far off and not be able to come aboard that night they would be in danger of freezing, it being winter at the time. The men took the boat and set off. The weather proved to be very fine until towards evening. About three o'clock, two of the men returned with a bountiful supply of eggs, about one hundred apiece. I asked them where the others were and they said that they had gone on up to the straits. Now going up to the straits is quite a walk, as it is some eight [illegible word] miles there. I told them that they were very sensible in not going, and they would find it so before morning. They said that they wanted to go but the others told them that they could not walk that distance and they sent them aboard with the eggs. "Mark my word," I told them, "you will be very glad that you did not go."

Night come on. It was just before dark when the first mate called my attention to three men on top of the mountain and presently we could see three more at the head of the bay. The wind had now shifted around to the northard, and it was coming on to rain. I told the first mate to lower away his boat and go and get the men at the head of the bay. He accordingly did so. I also told them [to] remain there and wait until the other three from the top of the mountain came down. He had not been gone a great while when he returned with the three men and said that he did not wish to remain as it was coming on thick and heavy. After supper I sent another boat off to the head of the bay. The rest of the crew went to Bull Beach thinking that they might have wandered over there in the darkness, as it was as dark as pitch; in fact you could not see your hand before you. These, like the former ones that went in quest of them, proved of no account, so they returned. At last I told Mr. Usher that we had better go to the rescue, so we had the boat lowered and with six men we put off. We had some difficulty in landing, but managed to do so after getting our feet wet. As I have said, it was a very nasty and unpleasant night. I had brought a lantern with me, thinking that it would be of some good to us, but this turned out to be the reverse, as you will see presently.

We had just landed and were walking along the beach, trying to feel our way through the darkness. You would skip and fall and then mire down into a mud hole, when the fellow with the lamp, a boatsteerer, ran afoul of a big bull elephant. He fell prostrate on the bull's back. I could not tell who was the most frightened, the elephant or the man. The lamp was dashed to the ground and naturally extinguished. The boatsteerer was in a rage. "Blast the infernal thing," he said, "instead of leading us on in the good path of righteousness it led us along in the path of destruction."

He then said something about not being [able] to see two yards ahead of you with it and that the blasted thing was not fit carrying. I thought so myself. It was useless to hunt for it as it was rather too dark and the ground was very broken thereabouts. After having proceeded up the beach for about a hundred yards, and not seeing nothing, all the time calling out at the height of our voices, I told the men that we had better go inland. So we started. Now the ground was more level and [we] were not in danger of breaking our necks by falling into some hidden crevice or hole. We had brought our rifles with us. I told the men to fire off a couple of shots. They did so. We kept up a regular fusillade, but could not get any response. We were now drenched to the skin. At last I told the men that we had better return to the boat and go aboard as the men could not be found and they would have to be left to their fate. It was their own fault and they should learn a lesson. We accordingly returned to the boat and went aboard. We found Mr. Joseph there. He had met with no better success than our party. It was being past twelve I gave them up for the night.

Next morning, a nice pleasant one, Mr. Usher came down into the cabin and told me that the three men I had seen on the previous day on the mountain had come down and were on the beach waiting for the boat to go off after them. I told him to lower away his boat and go and get them. After [his] having been gone but a few minutes, I went up on deck and could [see] the men coming off in the boat. They were a sorrowful looking set. They looked more dead than alive. After getting them aboard, I gave them a good reprimand by telling them that they should be ashamed of themselves for putting us to all the trouble they did, and I asked them to tell us how they had fared on the mountain the night before.

"I tell you the truth," spoke up one, "when I say that we never suffered so much in all our lives, as we did on that mountain the other night. You know yourself that it was bitter cold, and that we were in danger of getting frozen to death. Taking a short cut across the mountain, we had just reached the summit of it, and it came on very thick, in fact you could not see five feet ahead of you. We were afraid of descending from the mountain as we were in danger of being precipitated down some precipice, so we got to leeward of a projecting cliff. I knew that if we laid down, we would fall into a death slumber and never awake. Some of the men were not aware of this, coming from a tropical climate. They were Cape de Verde Portuguese, and some of them insisted upon lying down but we would keep rousing them up." I then asked them if they wished to go in quest of eggs again. Not much, they said, especially in this kind of weather. After having a good hearty breakfast, the principal article being eggs, they went ashore and brought off about four hundred eggs that they left up the mountain.

The next few days we were kept busy getting different things ready to go elephanting. One day three of the men come to me and told me that they were suffering from the scurvy. I was astonished at this as they had plenty of fresh and wholesome food to eat, and I told them that I could not understand how they could have contracted the disease. I asked some of the foremast hands about it and they told me that the two men in question were inclined to be very filthy, as they would come in to the fo'castle drenched to the skin, their clothes all besmeared with grease, and turn in with every stitch on.

I went and summoned the men and had them strip stark naked. I then had four of the men take scrubbing brushes and give them a good washing down. At first they resisted but I made them stand and take their bath. While this very rough ablution was

going on, I had Mr. Usher go ashore and procure some penguins'
eggs. When he came back, I had the cook bleed them. I then had
the men drink the blood. At first they were sick at the idea of
doing it, but I told them that they had to take their medicine.
There happened to be a German amongst the three and he refused
emphatically to take any of the gore. I told him if he refused to
take it that I would have the men to hold him and then force it
down his throat. He then commenced to cry, thinking I suppose
to gain my pity so that I would let him off. But this would not
work for I knew it was for his own good.

"I will never see my fatherland no more," he bawled out with
the strength of an Arizona jackass. "Mine Got in Heaven if I vas
to drink dat, I vas sure die," he said with the tears coursing down
his cheeks.

"You take that or by the gods I will carry my threat into
effect."

Seeing that I was in earnest, he grasped the cup of gore and
commenced to sip it, or at least to pretend to do so. "Drink it
down," I told him. "You put me [in] mind of some granny sipping
tea." I assure you that this poor fellow thought that it was a fatal
draught that was about to take. After contorting his face into all
kinds of ridiculous grimaces, which gave him more of the appearance of a monkey than of a Teutonic, he shut his eyes, thinking I
suppose that he would never open them, and gulped down the
gorish draught. I then had the steward skin the penguins and cut
the meat up and soak it in strong vinegar. [I] gave him special orders not to allow the men to have anything that contained an extra
quantity of salt. They were to eat this flesh raw, three times a day
until they were cured, and [the steward was] also to administer a
dose of blood [to them] at intervals.

In three days time they got so that they could drink blood and
eat the meat with zest. The German especially seemed to relish it.
I asked him how he liked it. "It was just taste like, vat you call him
in Inglish. . . ." I did not wait to hear the rest. He mumbled out
some big many-syllabled word in his native tongue. He was telling me about some dainty Dutch dish that tasted like penguin
blood. After four days they were as sound as ever. The disease

had completely disappeared. I then told them that if they would not keep clean and take off their clothes when retiring that I would take their clothes away from them and make them go around with a canvas bag about them and that if they would not wash, I would tow them over the side. These threats served the purpose for they ever afterwards kept clean and disrobed when retiring.

The blood of penguin, and in fact of any sea bird, is the best remedy for the scurvy that I know of.[30] You can get them in any abundance whether at sea or in any remote island. The best way to capture them at sea is by using an ordinary fish hook baited with a piece of salt meat. You can always manage to get sea hens, albatrosses, cape pigeons, whale birds, and many other kinds too numerous to be numerated here.

Several days after the above-mentioned occurrence, I gave the men orders that we were to go in quest of elephants next morning, so everything was got in readyance. Next morning found us on the beach slaughtering elephants right and left. That day we killed forty-four bulls, skinned them, cut the blubber off, and as we were not going to take it aboard immediately, we put it into water holes, so as to keep it cool and to keep the birds from devouring it. The heat melts the oil out of it and so much is lost.

In conveying the blubber to the water holes, poles are used, two men to a pole. On Bull Beach we use [a] cart but the rough ground here would not permit a cart being used; the above place was very level. It takes seven men to man the cart. These are their assigned places: one man in the shaft to steer, and six men to do the pulling, three on each side. The capacity of the cart is ten barrels of blubber. The first day we got aboard sixty barrels of blubber. The first day is the most awkward to the men, as they have to break themselves in, but after this everything goes on all right. After the men have got acquainted with their duty they can do much more work. We generally work from sunrise until sunset in good weather. A good spell of weather is not of very common occurrence and so we have to avail ourselves of a favorable spell. Our motto at all times is, "Make hay while the sun shines," and we generally do so.

Next day we went to Bull Beach, the adjoining beach, to kill

elephant; killed fifty bulls and the same number of cows. We
could not do much that day as it was raining and snowing very
heavy at times, and the most of the elephants were at the upper
part of the beach. Next day we killed the same number of bulls as
the previous day, and thirty cows. We now could make use of the
cart to great advantage. It did not take us long to convey it to the
water holes. The weather now increased in its inclementure. It
was snowing and raining very much and now commenced to blow
a living gale from the southard and westard. It blew so heavy that
a man could not stand up and skin, and as fast as they would skin
the sand would be blown onto the elephant, and in cutting the
blubber the knife would come in contact with the sand and make
them dull; in fact it was an impossibility to keep an edge on the
knife. The sand would be blown up into your face, blinding you.
Then the men would commence to rub their eyes and it gave them
a comical appearance. The grease, blood, and dirt made it an impossibility to tell whether they belonged to the Caucasian or Black
race.

I gave the men the order to stop skinning and to dispose of the
blubber by burying it in the sand, there being no water holes
thereabouts, to keep it away from the birds. The men worked
with a will and in a short time had disposed of all of it. We then
went aboard; it was about 2 o'clock. The wind was not near so
strong as there was ashore as our position was more sheltered. At
the mouth of the harbor the sea was breaking very heavy, but it
did not extend up to where we were. The wind went down
through the night, and the morning turned out to be a favorable
one, so we got ashore and commenced to work again. We killed
thirty bulls and as many cows, skinned them, but we did not get
the blubber carted up on account of the beach being so wet and
muddy the cart would mire down into the mud. So we buried
what remained in the sand. It bearing near night fall, we returned
aboard. Next morning proved to be fine, so we were ashore killing
and skinning again. As I have said we did not do any killing on the
previous day.[31] So at 2 o'clock we stopped killing and skinning
and went carting the blubber to the beach. We unearthed the
blubber that we had buried and treated it in like manner.

We experienced no difficulty in hauling the blubber from the beach as the northerly winds made it very smooth. But hauling blubber from a beach like Bull Beach is no easy matter. It is invariably rough on this beach and it is very dangerous work. The way the work is gone through with is by getting your blubber down to the beach. You have to carry it to the water's edge. You have two boatsteerers stationed here, whose duty is to take a rope about three fathoms long, with an eye in one end of it. They then take a separate piece of blubber and put the rope through the hole that has already been made in it, and they then take the end of the rope and put it through the eye. This serves as a barrier to keep the other pieces from coming off. You then get on all the blubber on to the rope that you can. After getting the first rope full you take another one and bind it on to it. You treat this and as many more as you think you can manage.

You then fix your raft rope straight along the beach. You generally haul off about thirteen raft ropes of blubber at a time. As soon as you have blubber enough the mate and his boat's crew launch the boat and pull for the vessel. They are equipped with a tub holding one hundred and fifty fathoms of line and a small kedging anchor and also twenty-five fathoms of fishing line fixed with a small weight on one end of the tow line. The boat then comes astern until it gets on an edge of an outside breaker, then heaves the fishing line on to the beach. The men ashore get a hold of it and heave the tow line ashore and make it fast to one end of the raft of blubber. As soon as the line is made fast the boat runs out and leaves loose all of the tow line, which has been made fast to the anchor that you have in the boat. As soon as you have got to the end of the tow line, let go the anchor and haul the boat back about fifty fathoms so as the boat can have full play and then there won't be any danger of the anchor coming back home.

While you are making these preparations, the men ashore are employed in getting the blubber out into the surf so that it will float. It needs the strength of every man in the boat to hold good and taut on the tow line, for if a constant strain is not kept on it, the surf will swamp the blubber up on to the beach, then we have to do all the work all over again, but if the men take precaution and perform their duty right, this danger can be averted. But nine times out of ten when the breakers recede they will take everything before them. In fact, it will make a clean sweep of the beach. When you perceive that the surf is going to take the raft out, you give the word to stand clear; the men obey by doing so.

But it happened the day in question, one of the [men] neglected to do so with promptness. The result was that he got entangled in the meshes of the raft. The raft was fast receding from the beach, and the poor fellow was going out with it. I could see that it was sure death for him if he got into the breakers, for there was a heavy surf coming in at the time. So I went to the rescue by springing in after him. I was just in time to grasp him by a shirt collar as he was going into the first breaker. As I got hold of him in his struggle, he grasped me by the arm. I just had time to get braced good when the next breaker come when it struck me. Oh my, but I felt a queer sensation all over my body. I felt like as if the whole Indian Ocean had rolled over me. As it struck me it capsized me completely. I made one complete somersault and the drowning man went a shooting up on the crest of the breaker like a rocket. The men could see [us] coming and they stood ready to grasp a hold of us when the surf would wash us in so as to prevent us from going out with the undertow. Up, up, we went on the crest of the wave, and away the wave carried us onto the beach. We struck terra firma and [were] taken a hold of by the men and hurried — the man had to be carried — to a place above high water mark.

Having got ashore O.K., the poor man looked actually dazed. I myself felt rather exhausted, as I had exerted myself a good deal. After allowing the man a spell to recover his breath and get back to himself again, I asked him how he enjoyed surf bathing. "Captain," he said — I could see gratitude streaming in his eyes — "that is the most dangerous and perilous moment that I have ever been subjected to. In fact, I thought that my time was up. I thank you many times, captain, for your kind assistance in rescuing me from a watery grave." I accepted his thanks.

I gave the third mate orders to be getting down another raft of blubber, as I was going aboard to change clothing and I would also take the man I had rescued with me, and I also told him that [if] he got the raft ready during my absence he could start and get it off. I took the boat and went aboard. During this time the rescued man was pouring out his gratitude to me. I gave him to understand that I thought it was the duty of one man to his fellow man. I did admit that it was a miraculous escape, but it pleased the Almighty to spare us, and so let us be thankful to Him. By the time that we got to where the vessel was, we were put aboard by the [mate], Mr. Usher, who said, "You and that man had one of the closest shaves that I have ever seen a man have, and captain, I must compliment you on your valor in rescuing that man, who would have surely met with a watery [grave] but for your timely rescue." I told him that I was in duty bound to have tried to save the man's life. And if I had not made an attempt to, I would have ever afterwards felt a remorse for it. We accordingly went aboard and changed our clothes, had a warm drink of coffee, and went ashore again.

Having arrived on the beach, I found that the third mate had the raft of blubber all in readyance to take off, so we get the line out. I told the men to stand clear and look out that they did not get foul of the raft. We managed to haul off forty-five raft ropes of blubber that day. The succeeding day was not so good for hauling on account of it being very rough out in the surf. So instead of rafting we went to killing sea elephant. We killed sixty bulls and about one hundred cows. The officers done the killing and the men were kept busy skinning, except two who were put to carrying blubber on the poles.

We worked rather late that evening. In fact, it was dark when I gave them the word to get ready to go aboard. I must say that it was quite a treat to us having worked so hard all day and we were literally besmeared with gore from the brutes. We got aboard and after a good hearty meal, the principal article being elephant heart with sage stuffing, which proved very palatable — in fact I prefer it to bullock heart if prepared right — we retired for the night and it pleased God to bring us to the light of another day, and the weather turned out to be very fine. It was now very fine on the beach. As we were breakfasting, I asked Mr. Usher how many raft lines we still had [to bring aboard]. He said about seventy. I told him that as it was a fine day for rafting he must try and manage to get about a hundred raft lines. After breakfast I gave him orders to hoist the blubber that we had alongside aboard and have it stowed away in the hold as quick as possible, as we were going to do justice to our motto that day: "Make hay while the sun shines."

"Aye, aye, sir," he said, "we will make that blubber disappear in a jiffy."

He was good to his word, for in about an hour he had it all stowed away. I then gave him orders for him and Mr. Joseph to procure all the raft lines aboard, and lower away those two boats, and pull down to the beach where we had been killing on the previous day. And I told him that I would follow suit and go down and get the poles and the rest of the raft lines. "All right, sir," he said, "and if you do not come down soon enough you will find us rafting." They lowered their boats away and pulled for the beach. Shortly afterwards, we [followed] them. Arriving there, I found the men hard at work rafting. Mr. Usher I perceived had carried [our] saying into effect. That day we rafted them off at twenty lines at a time. That day we hauled off one hundred and five lines "chuck a block full," as Jack says. I assure you that it was a good day's work. But as I have said, when the men get to understand their work, they can do three times as much as when they are not acquainted with their different duties. We got aboard that night feeling very tired. After a quick meal, this time of leopard seal, we retired for the night.

Next day proved to be a good one. Mr. Usher and fourteen men stayed aboard to try out, and I took twelve men and Mr. Joseph [the second mate]. We went ashore and commenced to kill. We managed to get ten bulls and thirty cows. We thought that this would be sufficient to keep us a'going, so we went on board early that evening and after partaking of a heavy repast of Desolation cabbage, we went to work helping the rest of the hands mince. At dark we let up mincing and boiling. We got fifty barrels of oil that day by working from 6 o'clock in the morning until 7 o'clock that evening.

Next morning at daylight, I got eight men and went ashore to finish up killing the elephant there was on that beach. In time we killed all the elephant, left about ten cows. I left these so as they could call up more bulls. We went aboard that evening feeling rather tired. After supper I commenced to sum up the number of elephant we had killed since starting to work this place, and I found that we had killed two hundred and twenty-three bulls and two hundred and thirty-two cows, and we had about half of the blubber on the beach. It is very difficult to cart it here, as the beach is one mile long. For three days we kept carting the blubber. On the third day it came on to blow from the northard. This made the beach smooth so I stopped all the other work and commenced to raft the blubber aboard. We got all the blubber that was down at the lower beach and one raft from the upper beach that day. It was very fortunate to us that we done this that night as [the] wind came around to the N.N.W. and stayed there for ten days. During these ten days we were engaged in trying out and stowing the oil away. I also kept a few men on the beach killing what few elephant that came up. They killed some twenty-six cows and bulls and carted the blubber to the beach.

A few days afterwards the weather again turned out fine. As we had finished boiling [the blubber] and had it all stowed down, we commenced to clean up and got ready to leave Young William's Bay. We had about six hundred barrels of oil. About two weeks afterwards the wind came on fine from the northeast. We got under way and got down as far as Cape Bourbon. The wind then dropped to a calm. It came on very thick and commenced to rain. We stopped out all night under short sail. At dawn the wind hauled around to the southwest, but was very light. At daybreak I kept away for Sprightly Bay. Shortly afterwards we came to anchor there.

We met with very good success here. We worked hard for about thirteen or fourteen days. And when I came to take an invoice of our cargo, I found we had 700 barrels of oil and fifty barrels of loose blubber. We were to take this to Three Island Harbor and try it out. I could not dispose of it here as we had no empty casks. On the fourteenth day the [wind] was blowing from the western. We got underway. As soon as we arrived on the outside, it commenced to blow a gale. We were in immediate danger of being blown ashore if we hugged the land. So I thought best to make a harbor. Swains Harbor being the handiest, we made for there. While going in, it blew so much that we had to shorten sail, so all sail was taken off but the jib. When we got to the mouth of Swains Bay, the wind was blowing very hard and at random, and also a tremendous sea on. I asked Mr. Usher, the first mate, what he thought of it.

"Well," he said, "captain, I don't think that there is any earthly show for us to get in there. If you would take my advice you would run out. We will get all the canvas blown off of her if we attempt to sail up the bay."

I was undaunted at this so I gave orders for the mainsail to be [reefed] and also the foresail with double reefs in each. For about twenty minutes' time these two sails were set. The wind kept blowing up the bay very strong. I gave orders for six men to stand by the main peak halyards. These men were to lower away the peak when a heavy gust hit her, and to hoist when it had passed over. No doubt but what it was a very perilous venture as we were in danger of being run ashore at any moment: we were at the mercy of the wind.

We commenced to work our way up the bay. The wind had continued blowing up the bay. We had to lower away the main peak as a heavy gust would come on and then it would lull down. We had made very good progress, and I was highly elated over it, when we got rather aback [and] the wind came on dead-a-head. This was rather unexpected as we thought that the wind had hauled around steady to the south. I immediately gave orders to the men to take in all sail. Before this, the first mate took soundings and found that there was seven fathoms of water and good holdings for the anchor. I gave orders for them to let go the large anchor and sixty fathoms of chain. I gave orders for every sail to be ready to hoist up at a minute's notice. The [sea] was heaving very much, and the wind was luffing, which made it worse. The wind would strike the vessel in the stern, which would cause her to shoot ahead. This created quite a strain on the anchor. The wind would then haul around to the other quarter and send her a shooting astern. The bay being situated between two very highly elevated pieces of land, the wind would come down between these two pieces of land in a proper gale. We lain here for about four hours.

Shortly after dinner I was on deck and indulging in a smoke. The weather did not look as if it was going to clear. Mr. Usher joined me; [we] were walking the quarter deck. The wind had not abated in the least, when I heard something give a loud snap. Looking forward, it was plain to us what had happened. The chain had snapped. I gave the order for all hands on deck and for them to hoist the jib and foresail. The men sprang to the halyards and in a short time they were up, and a'drawing. I sprang to the helm and put it hard a'port, but she would not pay off. She kept heading for the land. My God, we were surely in for it hard. I sang out for the men to be quick and hoist the mainsail and to be quick about it for our lives depended on it. It was now rather close as the jib boom was dashing into the breakers.

At last a strong puff came up the bay and hit the mainsail. This made her pay off-shore. The next gust luckily was from up the bay and it sent her headlong down the bay. Luckily it was the last heavy gust that parted the chain. The wind now commenced to grow more regular and lighter and so we went around and commenced to work up to Swain's Islands. We had to sail about twenty miles before we could come to the anchoring ground. We managed to get there at about 4 o'clock; we let go our anchor.

At about 5:30 one of the men came running aft, saying that there were three men on the opposite shore. I went forward and found out what he said to be true. I immediately had a boat lowered and went on shore and found the men in question to be officers of H.M.S. *Volage.*[32] They were very glad to meet me. They stated that they were lying at an anchor at Royal Sound. The *Volage* had just shortly arrived with the transit of Venus party. They had erected an observatory. This was situated about one mile from Swain. [We] hauled over. They asked me what schooner that was. I told them that it was the *Roswell King.* The senior officer knew me at once. "Captain Fuller," he said. "I am ever so glad to see you." I was rather astounded at this and I told him that he had the advantage of me by being acquainted with my name.

"It is true, captain," he continued with an amused look on his face, "that we have never met before. The way that I came to hear of you was through the reports of Captain Nares of H.M.S. *Challenger*. [33] He spoke very highly of you in his reports, that were submitted to the British government. I assure you that the captain of the *Volage* will be very glad to see you, as he has been mentioning your name in conjunction with going to Heards Island. We left a letter over at Three Island Harbor. I think the contents of it were that he desires you to come up to Carpenter's Harbor in the sound." [34]

They then commenced to relate a good deal of very interesting news to me, and then followed a lot of questions, concerning the island, and about elephanting, etc. I gave them the desired information in regard to elephanting, etc. They told me that there was a party of Americans in the sound on Malloy Point, and also a German party at Bessie Harbor (or better known as Pot Harbor). It derives the latter name from the enormous quantity of old try pots that have been left there. I then invited them aboard. They said that they would be only too glad to accept the invitation but, as it was close on to nightfall and it would be dark before they could get back ashore, they would decline the invitation and accept it at some more convenient period. "I assure you, captain," spoke up one, "that it would [be] quite a treat to us to go aboard of your vessel. I will make it a special point to do so." Having exacted a promise from me to come on board next day, we parted. They returned to the station and [we] went on board. [35]

Next day the wind was the same, still blowing very strong. After dinner Mr. Usher, his boat's crew, and myself went ashore and after a walk of 1½ miles we arrived at the station. I was received very courteously by the officers. They showed me through the observatory, explaining the use of different instruments, astronomical, that were used in taking observations. They also seemed to be deeply interested in different subjects concerning the island. I spent a very pleasant and interesting evening amongst them. It was sunset when I took my leave of them and I got aboard at dark.

Next morning the wind hauled around to northwest and still strong. We got underway at 4 o'clock and headed for Three Island Harbor; at 3 o'clock that evening we came to anchor there. After making everything snug, I had the [boat] lowered and I went ashore and to the house that is there and found two letters addressed to myself. One was from the owners of the *Roswell King* and the other from the captain of Her Majesty's Ship *Volage*. The contents of the one from the ship's owners was to the effect that they did not want the transit of Venus party to interfere with my mission in those parts, and that if I were to do anything for them, to charge them for it, as the English government could very well afford to pay something for valuable information. These words were meant no doubt to put me on my guard against the different parties, and I must say were uncalled for. I had come to the island for oil and not on scientific research. The letter from the captain of the *Volage* contained a request for me to come up to Carpenter's Harbor.

It being too late for us to commence work aboard that evening, I had part of the men go aboard and bring off my shotgun. I was going out gunning after duck, they being very abundant on this part of the island. They brought my gun off and, with one of the men to accompany me, I started off. Having a knowledge of my surroundings I knew where the game were. After transversing the beach for a while, we struck off into the foothills. It was not long before we scared up a brace of ducks. At last we arrived where their rookery was. The molting season had just sat in, and vast numbers of them could be seen sitting on their nests, tending to their young and acting as incubators. During this time of the year they are partially unable to fly; at least they can not soar in the air. You can approach them as near as you want to and they won't make any movement to escape. Of course it would not be sportsmanlike to take advantage of the birds by killing them on their nests, so I satisfied myself by killing the ones that were able to fly. After killing some fifteen braces of them, and being rather tired with the evening's sport we retracted our steps toward the beach.

As we were walking leisurely along the beach, we run upon a pair of sleeping [leopard] seal. Not having any irons with me so as I could kill them, I dispatched the man to the boat after some, as it was not very far. He shortly returned accompanied by the boat's crew. I took one of the irons and plunged it into the brute's shoulder. It was plain that it had touched a vital spot, for it gave a few struggles and expired. The other one in like manner was dispatched. I then gave orders for them to get knives, and a couple of poles, and dispose of them; I also gave them orders to save part of the flesh as it is very good eating and wholesome. It is considered a dainty on the island. The flesh it has the appearance of beef and tastes much like it only a bit stronger.

The leopard seal are not so plentiful as the sea elephant. They also taste a good deal like a hair seal. Their head resembles that of a sheep, only the nose protrudes out further than that of an elephant. Their tongue is very much like a bovine's; it is long and fitted with a rough surface. They have two small holes that serve for ears. Their jaws are much longer than those of an elephant; their teeth are different in respect to shape and size [and] are solid. The molar teeth are pronged and set very close together. Their color of their coat is of bluish hue and at times running into a steel blue or jet black. They are adorned with numerous spots, dainty in color and about the size of a half-dollar piece; these are very numerous on the back. [They] derive their name from these spots. The belly is of a white color, at times to be of a slatey white. The male and female attain the same size. They are about seven or eight feet in length and four or five feet around the stomach. A good sized one will produce three-quarters barrels. The oil is of the same quality as the elephant.

There are different ways of preparing the flesh for eating. Made into balls, it is delicious and also goes very well fried or made into stew. Of course the young ones are preferable for eating as the flesh is more tender and juicy. Large ships coming to these parts consider the leopard seal quite a delicacy. Every particle of the flesh is kept and cooked into different dishes.

After having given the men orders of how to dispose of the brutes, I went aboard. Shortly afterwards the men arrived with the boat. They had it laden down with the flesh and blubber of the leopard seals. It being too late to prepare any of the meat we reserved it for the morrow and had duck for supper. They are very good eating. In appearance they are like our teal duck at home and will average a pound apiece. Their shape is identically the same and their color differs but little. They have a small tuft of light steel blue feathers in the wings. They are very good eating and I think equal in flavor to any fowl. There is one peculiarity about their eggs. As I have said before, they build their nests on the bare damp ground. Well, the eggs have a very peculiar taste, liken to damp soil, which makes them unfit to eat, or at least not very palatable, although I don't believe they are unwholesome.

Our supper that night was comprised of baked duck, stewed Desolation cabbage (this is indigenous to this island) and mussel chowder tasting very much like Chesapeake oysters. These can be procured in any abundance and they are delicious when served up either stewed, fried, on shell or raw with vinegar, salt and pepper. The meal was quite a treat to us, and one [not] to be sneered at. After eating our full, we had a smoke, a walk on deck to help promote digestion, and we then retired for the night.

The next day the crew was busy boiling out the loose blubber we had aboard. We managed to dispose of it in one day. We put it ashore and then broke out some hundred barrels of oil and placed them ashore. Next day were engaged in cleaning the vessel. She looked very untidy for the handling she had gone through. It took us about a day to clean her up, and on the next morning, the wind being favorable, we got underway and stood off for Carpenter's Harbor.

On our way there I noticed a very interesting performance between a flock of sea gulls and a leopard seal. Being to windward of them and not very far off, we could witness the whole affair. At first one of the men drew our attention to it. The leopard was lying in water rolling about, as if he were either asleep or dead. A whole flock of gulls were hovering around him. Now and then one would fly down and pick at him. They then became more daring.

They would alight for an instant. At last like a flash he grasped one of the unwary birds in his jaws and disappeared. Ten or fifteen minutes afterwards, the skin of the gull came up to the surface of the water. I do not know if he repeated this ruse, as we had lost sight of him afterwards.

Arrived at Carpenter's Harbor. I could see that the English had taken it up as a stopping place, for I could see groups of men and officers walking up and down the beach. They had erected quite a substantial observatory. After getting things fixed aboard I had the boat lowered and I went aboard. I was met by Captain Fairfax at the gangway, who introduced himself and welcomed me aboard. After receiving an introduction to his officers, he invited me into the cabin. We exchanged the compliments of the day. He then said, "The men that met you over at the west station informed me about you. They told me that you had been in Swains Bay and that they had exacted a promise of you to come here. I must say that I am very glad to meet you. I had commenced to think that you would not come up here. I want to consult you in regard to my going to Heard Island. If you think it advisable I will put a party on there to take observations."

"To tell you the candid truth I don't think it would be safe for you to venture over there. In the first place you will be exposed directly to the wind and surf, and your vessel is rather too large a craft to be in close quarters. There is no shelter for you as properly speaking there is no lee side to the island. You will be obliged to keep up steam all of the time as it is not a place to anchor at, and again it might be [a] fortnight or so before you will be able to land the party. They would have to go in on the surf and their instruments would be in danger of getting ruined. My advice to you is not to go there. You had better consult Captain Bailey of the *Emma Jane* or Captain Sisson of the *Colgate*.[36] They are better acquainted about the island than I am."

He then told me that he asked the advice of the two mentioned gentlemen and they had advised him to go there. "They said that they did not think there was any danger in us going over there. But this does not cut any figure in the case at all. The British government gave me orders to be advised by you as Captain Nares

spoke of you as the best authority on questions of navigation concerning the island."

"You can do what you are a mind to, Captain Fairfax, but if
you take my advice you won't venture around that quarter."

"Your advice will be taken, Captain Fuller," he said; "I won't
send a party there. I will be led by your advice in the matter."
After a few more words in regard to different subjects concerning
the island, the captain passed around the Havanas. He told me
that I ought to go and see Father Percy at the station. He was the
head of the scientific case. "I want you to take dinner with me, so
be sure and be here at 6 o'clock sharp. Now don't fail to come, as
you will disappoint me [otherwise]."

I bade him good [bye] and went ashore, and met the captain of
the *Enterprise,* tender to the H.M.S. *Volage,* and Father Percy.[37]
The captain was one of those self-conceited English aristocrats,
[who regard] a master of a sailing vessel [as] of no importance. I
was not favorably impressed with him. Father received me very
courteously. I spent two or three hours in his company and he requested me to remain to dinner. Of course I had to decline the invitation as I had promised to go and eat with Captain Fairfax.
"Well, captain," he said, "I want you to come on shore and see me
before you leave in the morning." I thanked him very much for his
treatment toward me, and took leave of him.

I arrived aboard of the *Volage* a few minutes before the appointed hour. I was met by the officer of the deck and conducted
to the captain's cabin, where I was greeted by Captain Fairfax. He
welcomed me and then he escorted me into dinner. The table was
laden with all the delicacies of the season, or at least that were in
season in those parts. I was doing justice to my appetite when the
captain said, "I have killed a sheep for you, captain; it is the last
one we had, and I had it specially reserved as a gift to you. I am
almost certain that you will appreciate it, as I am aware that you
do not get fresh mutton in these parts very often."

I thanked him very much and he added that he had several different charts of the island and a book concerning the cruise of
Challenger in the South Seas. He said that they were sent me
through him by his government. I told him that I felt highly hon-

ored for the forethought exhibited by Captain Nares of the *Challenger*. I would be only too glad to accept the charts, etc., sent by the British government. I had often expressed a wish to Captain Nares to come in possession of some authentic chart relative to the island. He [Captain Fairfax] delivered a package to me, saying that it contained the charts and book.

"Captain," he said, "you must be endowed with an extra good memory to be able to remember all of the reefs and rocks in the vicinity of this island. The island itself is a very large one, and it abounds in reefs and rocks. You know every nook and corner about the island I dare say, and are acquainted with all of the bays."

I told him that there was not any thing wonderful about my memory, but my knowledge of the island could be accounted for very easily. From boyhood, I might say, I had been off and on about the island. In fact, the best part of my life had [been] spent in these regions. I had been promoted from "green hand" to "boatsteerer" and so on until I was the master of a vessel. I was at home when cruising about the island. A knowledge of my native town could be set as a fit comparison for my knowledge of the island. I was acquainted with every street and block at home; I knew where every store or place of importance was situated. I could find my way at daytime or nighttime. The same way with the island. The only difficulty I experienced was in foggy weather. Then it was all guess work. At times I would forget the course of some harbor, but I would always generally come out right, although I must say I get in to rather tight corners at times.

"I must say, captain," he broke in, "that you must have a thorough knowledge of the place. But haven't you had some pretty close calls in your time, while cruising about the island?"

"Yes," I replied. "I will narrate a very perilous and almost miraculous escape tonight, if time permits."

"Yes, yes, yes," they all replied, "plenty of time yet, captain, not 10 o'clock. Pass around the Havanas." I accepted one, lit it, and proceeded to relate my experience.

It was during the latter part of last winter. We had been cruising up around Pot Harbor in quest of right whale. One morning, a very unpleasant one, at about 7 o'clock — we were just off of Kent Rock — when the lookout was heard to sing out, "There she blows."

"Where away there?" I cried out to him.

"Dead ahead," he replied. A few minutes afterwards, we could see the whale, not a great distance off from the ship. He would come to the surface of the water and spout. Both boats were lowered, each one with their complement of six men. We being to windward of him, the boats did not have much maneuvering to do, but still he led them a good chase. The first mate's boat was first in getting an iron into him. He turned flukes, and came up, spouting along. I will here add that it is a very known thing amongst whalemen, that if a whale does not spout blood when he has been struck, he is in great danger of being lost by sinking. In a short time he expired. Both boats got on to him and brought him alongside: we made him secure by getting a fluke rope around his flukes and then he was made fast and secure and we stood off for Pot Harbor.

It now come on very thick and the wind increased in strength. I could see nothing ahead. There is some very dangerous rocks around Pot Harbor, so it was a rather risky venture for me to attempt to get there during this kind of weather. I got to thinking that we must be very near land, although there were no indications to make me believe it. At any rate, I was under the impression that the land was handy. I sent for my second mate, Mr. Joseph, and gave him orders to lower his boat and to pull ahead of the vessel and keep a good lookout for the rocks or beach and to keep in sight. I then had the men attach a tow line to the whale's flukes, his tail, and to have an anchor bended on to this. The whale was. . . .[38] The second mate had been down but ten or fifteen minutes when I perceived him coming back. He was waving his hand and seemed to be in very much excitement. I knew what was coming. I sang out to the first mate to let go the whale. I then had the men haul aft the mainsheet. The whale having been let go, the vessel was more active and free in the water. She gave one terrific dive into the sea, and it was now plain to me that we were on the outer breaker.

I had the wheel put hard down and could see that if she would not come about we could not escape from being driven ashore. The mainsail having been lowered down, it was a difficult matter to come about. In fact, I did not expect she would. I have been on vessels that were said to be her superior as regards sea boats, and they would not come about when placed in a fix like we were, the mainsail being down. The danger come on so sudden that we were unable to hoist in time, so I took the chance and then like as if by magic I could see her coming around. We then run off on the other tack and stood clear of the land. You can hardly realize the immediate danger we escaped. I didn't myself until it was over and then I looked back with a shudder. I will say that this was one of my most narrow escapes. A couple of ships lengths further and we would have been back upon the cruel and merciless beach.

"A thrilling and narrow escape," said Captain Fairfax.

"But what ever became of the whale, captain? He must have been [a] gigantic big fellow," asked one of the lieutenants, who seemed to have evinced a good deal of genuine interest in my narrative.

"It calmed down about the fourth day. We got into Pot Harbor, and I then sent the boats after the monster. They hove up the dragging anchor, and both boats fastened on to him and towed him into harbor; [he] brought us about one hundred fifteen barrels of oil and one thousand barrels of whale bone.

"How about ambergris, captain?" asked the very muchly interested lieutenant.

"Ambergris — that is something that you hardly ever come into contact with. In fact, you might kill ten thousand whales and never get any, and of course [there is] a possibility of you[r] finding it in the first whale you catch. This time old Dame Fortune never smiled on me, for we never got any."

"What is this ambergris, captain?" one of them asked.

"It is a peculiar substance resembling beeswax that forms in a whale's stomach, caused by indigestion, or, in other words, costiveness.[39] A whale ailing from this complaint will oftimes float on the surface of the water. Hello; it is half-past [eleven]. I must be getting aboard, so you will excuse me."

"You have given us quite an entertainment with your narration, captain, and I hope it won't be the last one I hear from you." They then bade me good night. I then went to the boat. I must say that I was surprised when I got in the boat, as it was loaded above the gunnels with different stuffs, and the steward kept bringing more. I told him hold on or you'll sink her, but he insisted upon discharging his load. I thanked the captain very much for his liberality, and then told the crew to pull off.

Drawing alongside of the vessel everything was put aboard and I then took an invoice of what I had brought with me. First of all one whole sheep, quarter of beef, two doz. bottles of best English ale, half-doz. bottles of Port wine, ditto the same of Cognac brandy, and five bottles of Pomeroy Sec Superior Champagne. I must say that I now had a bountiful supply of choice liquors. I did not appreciate the liquor as much as the meats. I also got a lot of miscellaneous stuff, such as potted ham, sweet crackers, choice pickles, etc., etc., etc. After disposing of the meat, liquor, etc., I retired for the night.

Next morning I went aboard of the *Volage* to bid Captain Fairfax and the acquaintances that I had made aboard good-bye. I wished the captain a "bon voyage" and he did the same to me and he said, "I hope your voyage will be a successful one and that you won't have any more such close hauls with whale, as you did off of Pot Harbor." I told him that I did not know but what I would, but I hoped not. I then went ashore to take leave of Father Percy, as he had requested [me] to come and see him before leaving. I was met by him at the entrance to the observatory.

"Are you going to leave us this morning, captain?" he asked. I told him that I was.

"And where will you go from here?" he continued. I replied that I was going cruising around the island in quest of whale.

"Captain, your information relative to the island has proved of the utmost importance to us, as it has removed a good deal of inconvenience off of our shoulders. You deserve credit for your knowledge of the island place. And you really deserve a recompense of some kind. How long do you intend to remain in these parts?" I told him till the vessels come from Heard Island.

"Well, I wish you good speed, and may God guard over you and bring you back to your native clime." I thanked him very much for his kind wishes and told him I wished him the same.

"I want you to accept this box of me. You will find what the contents are when you open it aboard of your vessel."

I thanked him very much for his generosity and told him that I was only too sorry that I could not reciprocate. "Never mind, captain," he said, "You have more than reciprocated already by giving us information of extreme value and importance." The box was a good sized [one] and I perceived it was rather heavy when the men picked it up to carry to the boat.

"Well, Father Percy, good-bye to you and may you achieve what the height of your ambition hungers for."

He grasped me by the hand and said, "May you have a pleasant stay and run home." We then parted. Father Percy I must say was a very kind and congenial man. He was considered to be the smartest man in the post, and, in fact, was classed among the big heads of England. I can't but help of remembering of him with kind regards. I walked to the beach, got into the boat, and was pulled aboard. We hove up anchor and got under way.

Next morning I opened the box that I had received from Father Percy. I found there in a quarter of mutton, one doz. half-bottles of wine, and three large loaves of bread aggregating in weight about 10 lbs. apiece. The mutton came in very handy; the wine I put aside for future purposes; and the bread — what could of good old Father Percy been thinking of by putting bread into the box? I suppose he was under the impression that we lived on hard biscuit[s] at the time, but we had fresh bread baked every morning. I sent the most of it fo'ward to the foremast hands, who made it disappear very quickly. We sailed along before a good stiff breeze that blew across our quarter and a little before sunset we ran into Malloy Point.

The American Station being situated here, I went ashore and was welcomed by the chief personalities of the party, Captain Rien, Captain Train, and Doctor Kidder.[40] I spent a very pleasant evening with the party. I found them in very comfortable quarters. They showed me the numerous astronomical instruments they had for taking the transit with. I delivered a letter to Doctor Kidder that had been entrusted to me by Father Percy. At about 7 o'clock P.M. I went aboard. Next morning we got underway and stood off for Pot Harbor. I got there about midnight. It was a very unpleasant night, as it was pitch dark and raining. We let go our anchor, made everything snug, and retired for the night.

I was awakened next morning by someone hailing the vessel, but it was in a foreign tongue; I couldn't understand them. After breakfast, I told Mr. Usher that we might as well go ashore and have a glance around. The boat was lowered and went. We were accosted by a party of gentlemen donned in naval costume; a first glance told me that they [were] officers of a German man-o'-war.[41] I was accosted by one of the juniors of the party, who said, "Vat vas de mater mit the captan of your vessel, vas he no got any respect for the German government? Why he no show the respect due us by the flag? Vas you not see hoisten the German flag?"

I must say that I was rather abashed at this outburst of language or rather reprimand from one who had no right to do it. So I told him that I was not aware that there was any one ashore in this part of the island that wanted to know what flag the vessel was flying, but it was the *Roswell King*, Captain Fuller.

"Wen vill Captain Fuller come ashore?"

"I don't know but what he is ashore now."

"Vell, where is he?"

"I guess I am him."

The propounder of these audacious questions was then all honey to me. "How vas you do, Capting Fuller, how vas you do Capting Fuller? I vas ever so clad to make your acquaintance. Come the house right in. You vas velcome."

"No," I answered him, "there is plenty of room aboard of the *King* for me. I decline your hospitality."

"Captain, I hope you have not taken offense at what that officer there has said to you. You will excuse him as he was not aware that you were Captain Fuller. You will confer a favor upon me and the captain of the *Gazelle* by coming over to the station."

I looked around and found myself in [the] presence of a distinguished looking individual, about the average height, [with]

scanty locks of hair, strewn over his very broad forehead, fringed
here and there with grey. One of the officers that was standing by
introduced him to me as the chief professor of the scientific core of
the German transit of Venus party. He was a man of N. polar region fame. You will excuse me for his name, as I do not recollect
it. He then said, "I am exceedingly glad to have met you. I have
been very anxious to see you in regard to different subjects pertaining to the island that I want you to inform me on. You will
kindly step up to the house, if it is convenient to you." Seeing that
they meant to treat me with courtesy, I assented by saying I
would. The loud-mouthed individual seemed to look rather
sheepish, seeing the course things had taken since they were
aware that I was Captain Fuller of the *Roswell King* and not a petty
officer as he thought.

I accompanied the party up to the house, a very commodious
one, fashioned out of pine lumber. It was built in a very strong
and durable manner. In fact I think that they took too much pains
on it, as the beams of the house were about three [times] as thick
as what there was any need of them being. I was afterwards told
that they were under [the] impression that the wind blew ter-
rifically in those parts at times, so they took the extra precautions
to be fixed. The congenial professor continued by saying that he
wanted me to give him some information in regard to the sea
elephant, leopard seals, and birds that inhabited the island. I replied by saying that he was welcome to all I know of the place and
that I would give him the desired information to my full knowledge. "How did you ever come to be acquainted with my name,
professor?" I ventured to ask him.

"I can explain that to you very easily, captain. At first we
heard of you by reading the reports of H.M.S. *Challenger*. The
commander of her, Captain Nares, spoke very highly of you. He
makes mention of you in his reports, as about the most reliable authority on subjects pertaining to the islands that could be found.
Your name appeared in the English, German, and American papers. So it was a very easy matter to recognize you at a first
glance."

"Where is the German frigate, professor?" I asked.

"Over at Christmas Harbor, captain; I hope you won't leave without seeing the commander·of her; he has expressed a wish to see you. Perhaps you was not aware that he is wed to an American lady." I told him that I was not. And now followed a lot of questions from the congenial professor. In fact, the latter part of the day was spent in answering questions relative to birds, seals, sea elephants, geological formations on the island, etc.

I was introduced into the observatory. They had a telescope there of elephantine size and I was let look through it. Being a very clear night and the moon being at its best, I got a splendid view of old luna. You will excuse me for not giving you a description of what was presented to view, not being versed in astronomy. Suffice to say, it resembled a huge illuminated glass globe with dark looking patches on it. It is wonderful, thought I, 43 millions of miles away, and when the telescope comes to bear on it, it has the appearance of not being over 100 miles distant. They explained the use of different instruments. I spent a very enjoyable and interesting evening with them. The time piece was just sounding the midnight hour when I went aboard. That night I retired feeling rather fatigued, having continually through the day been absorbed in very interesting subjects, too much so to notice my weary feelings.

Next morning arrived and it turned out to be a fine one, accompanied with a fair breeze from the northard. We got underway at daybreak and ran down to Fuller's Harbor. In the evening, accompanied by two [of] the men, I started out in quest of some eatables for supper, as cabbage is very abundant in this part of the island. I had one of the men bring a sack with him to carry it in. We had not to go far before we came upon a patch. We plucked [a] dozen of longish and most solid heads and put them into the sack. The sack was now quite weighty as these cabbages when in a prime state attain a good size. They resemble our cabbage partly in shape, but have a different flavor, tasting very much like spinach. The inner leaves are white and very tender. They grow profusely about the island and of course are of a spontaneous growth. They are indigenous to this island. They are not to be found to the northard of 37°. We managed to scare up a couple of

braces of duck. One of the men came to me with a bunch of green
leaves in his hand and asked me what it was. Upon looking at it I
found it was what we call Desolation Tea, a kind of rose. It makes
a nice drink and is a good substitute for tea. Of course regular tea
is preferable. I had him pick a small quantity of it and put it with
the cabbage.

Having shot four more braces of duck, we started to retrace
our steps. On our route home we came across a rookery of penguins, known as "rock hoppers," from the fact that they build
their nests under rocks and wherever they can secret themselves.
We went over to where they were and commenced to look for
eggs. There was a large boulder projecting out of the ground. I
seen a number of penguins take refuge under it, so I told one of
the men to go and see if there were any eggs there. He accordingly
obeyed. There was an aperture large enough to admit his body, so
he crawled underneath the rock. The whole of his body disappeared underneath the boulder. He [had] not been there long before I could hear him polluting the air with some very blasphe-
mous epithets and I could see he was scrambling out in haste, in
fact quicker than he went [in]. I could hear the fluttering and flapping of wings, for [at] the time I was under the impression that he
was hauling out a couple of penguins by the legs. So I called out to
him to desist. I got no answer. All I could hear was the fluttering
of wings and the fellow swearing and yelling out. "Leave go, you
infernal old buzzard! Oh, I will fix you when I get you outside!
Leave go! Holy Smoke! Oh my nose!" and he came up scrambling backwards with a good sized penguin dangling to his nose,
and both hands full of eggs. "Here," he said, "captain, take these
eggs quick and let me ring . . . ," but before he had time to finish
the sentence, the penguin that was dangling from his nose let loose
and fled back into the hole.

Now the man was wild because he had not taken his vengeance on the altogether bold bird. He said he was going to annihi-
late the whole rookery. I told him he could go back into the hole,
and get the specimen out, and vent his wrath on it. He was afraid
to do this, as he did not care about venturing in there again and
take chances of having [his] eyes pierced out. So he had to satisfy
himself by thinking that he was instrumental in causing a dozen

and a half penguins not coming into existence. His eyes were bloodshot, and his nose was of purple hue, and about half an inch longer which give him the appearance of a Moses [or] Israelite.

I told him the occurrence reminded me of one of Mother Goose's nursery tales, about the four and twenty black birds. "The maid was in the garden hanging up the clothes, and down came a black bird, and bit off her nose." It was true that the penguins never bit off his nose, but as it was he would be able to lend this unfortunate maid enough nose to replace hers. He then wanted to take my gun and kill every bird in the rookery. I made him desist from carrying out his fiendish plot, and we made our way to the beach. We arrived opposite the vessel and I sang out, "Boat ahoy, there," and the third mate and his boat's crew came off after us. We had a very substantial meal of boiled cabbage and beef and baked duck. You can eat the cabbage either raw or cooked, and I have an idea that it would make excellent *kraut*.

Next morning the wind did not prove very favorable. At any rate I had the anchor hove up and we worked up to Breakwater Bay.[42] Arrived there and I went ashore and got some casks of blubber that I had left there. We also killed a few elephant that were about there. We then returned to Fuller's Harbor, [killing] what stray elephant we found. We then took and minced the blubber and put it in casks. We laid over in Fuller's Harbor to make preparations to go to Swain's Island in quest of seal. Several days afterwards the wind being from the western, we got under-way and worked to the outside of the foreland and the wind fell to a calm. We were making but little headway. Once in a while we would get a puff of air, only to die away again.

At about 9 o'clock A.M., we sighted a steamship. She was coming down from Christmas Harbor. She came up with us in about an hour's time. She was flying the American ensign and hailed us with, "What schooner is that?"

"The schooner *Roswell King* of New London," I answered. "What steamer is that?" I asked.

"The United States Steamer *Monongahela*, Captain Thornton," they replied.[43] "Is the American party at Pot Harbor?" they asked.

"No, the American party is in Royal Sound."

"Are you sure that they are not in Pot Harbor?" they again asked.

"Yes, I am certain of it as I was in Pot Harbor some four days ago, and the sole occupants of that place were the Germans. The Americans are in Royal Sound, at Point Malloy."

"What are you doing out here, and where are you bound to?"

"I am whaling and sea elephanting and am going over to Swain's Island."

They again asked, "Are you sure that the American party are at Malloy Point?"

"Most assuredly I am. You will be sure to run upon them in Royal Sound."

"Is there any danger in going down this passage?"

"Not a particle if you keep away from the kelp — sea weed."

"Many thanks to you, captain, and good-bye."

"You're welcome," I replied, and the *Monongahela* steamed away.

It now was a dead calm, so I sent away all three boats to Terror Reef after seal. They returned at noon with only three seal. I must say that they were chagrined at this, as I could see from their appearance. "Only three, sir," one spoke up in a kind of a meditative mood, "and we were expecting to get five or six hundred of them."

"Don't get discouraged, men, you will go to Swain's Island after dinner, and you are more liable to encounter there than in the place you have just tried."

Several hours after dinner, it still being calm, the boats were lowered away and they went ashore at Swain's Island. At 5:30 P.M. they returned having only captured seven seal. I must say now I myself was disappointed and the men were no less disgusted and put out about it. We had made all of the preparations for a good catch of seal. The men weeks before that had been hewing out seal clubs to kill them with, which was no small amount of work to make them. They had gone to the trouble to fix footwear for the occasion, which consisted in taking a very thick woolen sock and sewing [canvas?] to the bottom of them for soles. There is no danger of slipping with [them]; when they come in contact with rock they will adhere to it. And now they had a good opportunity to use them. Of course they had got nine or ten but not enough to pay them for the trouble. We had entertained high hopes of getting a good few, but we had been sorely disappointed.

This group of islands was formerly a noted place for seal. In fact, they were as abundant as what they are in the Arctics, but they have been gradually exterminated. It is supposed that upwards of three million and a half of seal have been taken off these same islands. Years ago the English worked these rookeries to great success. The skins did not bring such a good price in the European and American markets as what they do today. The best market to be found for them was the Oriental market. They would take them to China or Japan and get teas, silks, fruits, etc., in exchange for them. This I believe was a very profitable business.

A few words here in regard to how they are killed I am sure won't be amiss. The men go ashore, as I have already stated, armed with a club about four feet long. They also carry knives to skin them with. Of course they generally are acquainted with where the rookery is. Now, in this case of ours, they had no rookery. [Normally], they make straight way for the rookery. The [seals] that are nearest the beach are the first to be dispatched so as there won't be any danger of them escaping. If the seal starts to make for the water, the man confronts him and delivers a blow right on the nose. A well hit blow will generally kill them. He then bleeds [the seal], makes an incision with his skinning knife, and then proceeds to skin him. There is not much danger incurred in killing them, although they will protect their young and the clapmache[44] will often show fight. The yearling dogs are the ones that are looked for most eagerly, as they have the best fur. The whigs — mate's — clapmache is not so fine. In the breeding season they congregate in vast numbers on the beach and make a rookery, where they remain and rear their young, and then leave to come back next season. These supposed rookeries that we were looking for must have been broken up. They had been visited so often that they took fright and fled. It is a very singular thing how seal will immigrate. Now about to the southard of below 40°, barnacles do not exist. In fact, they cannot be found in cold climates. Now when these seal come up either to have their young or to shed their hair, they have barnacles on their bodies. They must immigrate to warmer latitudes, or again it might be that there exists a warmer current on the bottom of the sea where these barnacles can exist. I myself have not found out how it is. Suffice to say that I have conclusive proof of them going from one island to another.

Seeing that there were no more seal to be got, I had the men go ashore and kill a bull elephant. I myself accompanied them, and with their aid I managed to skin it through without injuring a bit of the hide. I then had them kill another one for its skeleton. Doctor Kidder had asked me if I would procure him a skin and skeleton so I complied with his wish. After getting the skin and skeleton aboard, we stood off from Fuller's Harbor, where we laid for the night.

That evening some of the men were bewailing their ill luck in not procuring any seal. Mr. Joseph then related an incident of his voyage before this to us that I must say made us almost turn green with envy at the herd their boat's crews could have availed themselves of. It seems that they went ashore on Hay Island, one of the Crozets. There were thousands of seals congregated here and it happened that the men that composed their crew were all "green" Portuguese. They advanced towards their victims. The seal seen them a'coming and of course started to make for the water. The Portuguese seeing this, thinking that they meant fight, dropped their clubs and fled instead of coolly knocking them in the head. They lost about five or six hundred seals by the cowardice of their men; if they had stood their ground, they could have got every one of them. I must say that my men were composed of this element as very near all of them had had previous experience in the business.

Hove up anchor next morning and made for the Port Palliser Islands. We were to work these for sea elephant. We met with but little success. That night we came to anchor in Exile Harbor. Next day there was a strong breeze from the northard, so we got

underway and run into Shoal Water Bay and let go anchor. Next
morning it turned out to be a pleasant one; it was breezing up
fresh from the westard. I told the first mate that I was going to go
ashore and then I was going to walk over to Boot Leg Bill.[45] The
boat was accordingly lowered and we went ashore. I started out;
five of the boat's crew accompanied me.

We were going to get some calico stones, a stone similar to an
agate, resembling a piece of calico very much on account of the
streaks that run through it. It was rather hard walking and we had
a good deal to contend with. It was some fourteen miles from the
beach to it. The route there was a very rough one, as you have to
climb up and down hills and go up and down valleys. On my way
there I picked up many curious formations of stone, several pieces
of crystal known as Desolation diamonds, from the fact that they
glitter and are diamond shaped although I don't think that they
are of much value. At last we arrived to where we were to get the
calico stones. There was any abundance of them, so we had but
little trouble in getting a supply of them. We then started to retrace our steps. Nothing of any interest happened during our
walk back. Now and again we would come upon a flock of duck,
but as we had not brought a gun with us, we could not procure
any.

We came upon a bunch of elephant as we neared the beach. An
old bull had about twenty or thirty cows encircled around him,
and the envious eyes of the young outside bulls were bent upon
him. Now and again one of them would venture to approach the
ring, only to retreat again, upon hearing the fierce warning of the
old bull who had the whole pod to himself. He would warn them
away by standing bolt upright on his hind flippers and uttering an
unearthly yell, sounding very much like a trumpet. At last an
extra bold one, of about the same size as the fellow in the ring,
tried to force his way in. He received a very warm reception from
the old pod keeper and then issued a regular combat. Their very
loud trumpeting and the very desperate way that they attacked
each other made them truly look hideous. Now one would rise on
his tail flippers and come down on his opponent, and then they

would commence to fight each other. They would repeat the act of raising on their flippers again; this time both of them would do it.

At last we could see too plain that the newcomer was going to be victorious. After a few more very desperate passes at each other, the old pod holder had to retire, looking much the worse for the fight as his shoulders were all lacerated and his whole anatomy in a bruised condition. Now the newcomer, looking as proud as a peacock, takes possession and of course he will have to retire from the ring if got the best of by some other aspiring young fellow. The old pod keeper now acts very affectionate towards the cows, as he will approach them and offer his caresses to each of them in turn. When a cow leaves the pod she makes straightway for the water. Every bull that she comes in contact with stops her and goes through a regular form of telling her good-bye.

We now approached very near the ring, but they did not seem to take any notice of us, they being so engrossed with their own proceedings. One of the men took and gave the old bull a prick with a pointed stick he had. The old fellow was truly enraged at this, for he looked around with a look of malignity as much to say, "Oh, you little insignificant fellow, if I only had you in my clutches, I would show you a bit of strength." After amusing ourselves as much as we wanted to, we resumed our journey to the boat. Arrived there, we hurried aboard. We were all broken up after the tramp. Some of the men complained of blisters on their feet. My own were feeling very sore. We all got supper and retired for the night, to enjoy well-earned sleep. I can't say but what we were fully recompensed for our walk to Boot Leg Bill as the stones we got were very nice ones as I have said similar to an agate and acceptable of a very fine polish, which makes them suitable for settings and various other purposes.

Next morning bright and early and immediately after breakfast, we hove up anchor and stood away for Three Island Harbor. We were to go in here to boil out the loose blubber we had aboard. We arrived there and started to work. Upon arriving there we found the U.S.S. *Monongahela* was at anchor there. We had got the boiling-out process fairly started when one of the *Mononga-*

bela's boats came alongside, in charge of the second and third lieutenants. The first lieutenant, Mr. Ludlow, introduced himself and the [second] officer, Mr. Strong, and [the] third officer, Mr. Miller.

They came aboard and commenced to inspect the vessel. They all seemed to be interested with the vessel, Mr. Strong, especially, as I am sure he had never been aboard of a whaler before in all his life. The inquisitive Mr. Strong was the first to commence to propound questions to me. We had two try pots a'going. Mr. Strong went over and commenced to look at them. "Why, you don't carry your galley — kitchen — out there in all kinds of weather, do you, captain?" He then stood on his toes and looked into one of the pots. He said, "Liver, eh? Is that the way you cook liver on this kind of packet?" This, of course, convulsed the men with laughter. I then explained to him that it was not our galley he was inspecting but the try works. I then pointed out the galley to him.

"By Jupiter," he said, "I was under the impression that that was your chicken coop. You will have to excuse my ignorance, captain. Don't think for a minute that I was [criticizing] your vessel." I accepted his apology, and he continued to propound questions. "So you don't get the oil out of the liver of the elephant?" No, I told him, as their livers were very much like a bullock's or other quadruped. They contained no oil, and what he had seen boiling in the pots was blubber.

"Where do you get this blubber from, captain, out of the inside of the elephant or on the outside?" I then explained to him and his companions the skinning, etc., of an elephant. "How much oil do you get from an elephant?" he again interrogated.

"According to the conditions and size of the creature we get all the way from half a barrel to seven barrels. That is the average."

"What is the most oil you have ever had an elephant produce?"

"Ten barrels, and of course he was an extra large fellow, in fact, about the largest that I have ever seen. I have heard of them getting more, and quite believe it."

"Do you take nothing but the oil of the elephant?" asked Mr. Ludlow.

"No," I replied, "we get whale and in fact anything that would make oil."

"What kind of a cargo have you to take home this voyage?" he again asked.

"I have got 1,600 barrels of oil, 2,000 lbs. of whale bone, and 100 seal skins."

"And how long have you been out from home?"

"Fourteen months."

"Do you consider this a good voyage, captain?" chimed in Mr. Miller.

"Yes, it was doing quite well, but everything depended upon the market. If there was no demand for oil, bone, or skins, of course, it would affect us, and you bear in mind that our expenses amount to a good deal." After they had satisfied their curiosity they bade us good-bye, at the same time offering me an invitation to dine with them that evening. They said that Captain Thornton had told them to invite me.

As they got into the boat, I noticed a familiar face among the men that were in the boat. I came to find out that it was one of the men that was in the schooner *Franklin* with me on my first voyage. Evincing some interest in the man, I asked Lieutenant Ludlow about him. "If your crew was composed of such articles as that fellow you must have had a good deal to put up with. He is one of the most worthless characters that I have ever came in contact [with]." I told him no, that the whale ship's crew was not made out of the same material as that fellow was; he was one of the most worthless men we [had] had aboard. We sent him home on the second season on account of his worthlessness and misbehavior. He seemed to be inclined to quarrel and there was not a man before the mast but what he had a falling out with.

"I can assure you, captain, that he won't try any of his little games aboard of our craft. Now, be sure and come over to dinner," said Lieutenant Ludlow. They then left and went aboard.

That afternoon at about 2:30 o'clock, I went aboard of the *Monongahela*. I was received very cordially by Captain Thornton, who invited me into the cabin where, after a pleasant talk on different subjects, I was shown into the mess room where I partook

of a very sumptuous meal. During the meal the captain asked me
if there was not a place in the vicinity of the island where he could
get good fresh water, as the water they were then using had a very
objectionable taste. I told him that I was aware of it myself that
the water was not of the best quality. I informed him that I knew
where there could be found a good supply of water. So before
parting we made preliminary arrangments to go to it on the morrow.

Next morning shortly after daybreak I went aboard of the
Monongahela. I was asked to breakfast but declined the invitation
on the plea that I had already breakfasted. After accepting a few
refreshments, the steam launch was got in readyance to carry us to
where there was water. We went aboard and steamed away to
Snug Harbor.[46] On the way there Captain Thornton and
Leiutenant Ludlow were asking me a great many questions concerning the island. Captain Thornton asked me if I did not occasionally have trouble with my men, he knowing full well the kind
of men that whalemen's crews are generally composed of. "I suppose," he said, "that you have some queer characters to comprise
your crews."

"Yes, Captain Thornton. I must say that we do. Ex-convicts,
jailbirds, and refugees from the law are as a rule found in the
crews of whalemen. I myself had a very close call to assassination.
I will give you the particulars of the occurrence, if you wish, as I
think we have ample time."

"By all means do, captain," spoke up Captain Thornton. Seat-
ing myself, and the captain and lieutenant followed suit, I commenced my recital.

"What I am about to narrate to you happened during one of
my previous voyages to this place. We were lying at an anchor at
Marienne Straits. One day we were getting elephant. We had
rafted a good deal of it off to the vessel and had a large quantity
alongside. The day proved to be a very windy one, as there was a
strong breeze prevailing from the N. western. We had all gone
ashore but a couple of men who were kept aboard to look after
things. It came on to blow very heavy, so we all went aboard.
Upon arriving there, one of the men come and told me that a

couple of lines of blubber had got adrift during our absence, and floated towards the beach. I informed the first and second mates of this and told them to go and reclaim it. Both boats were accordingly lowered away, and they went in quest of it. I watched the boats from the ship's deck. The second mate made his way up the straits. I at last seen the first mate's boat making the beach. There was quite a sea on, so they had to ride the surf in. They landed on *terra firma* and they then started to haul the boat up on to the beach. I noticed the mate ejaculating to the men and I then seen him lift his arm to strike one of them. Whether he did or not or whether he meant to strike the man, I cannot say. Shortly afterwards he returned with both lines of blubber and the second mate arrived shortly afterwards, and of course did not bring any blubber with him. After partaking of a hearty meal we retired for the night.

"As on all whalers, the mates, boatsteerers, steward, and carpenter lodge in the aft cabin. It had been my custom to leave the door of my berth open, so as I could hear when it came on to blow. After retiring I laid there in a kind of semiconscious state. How long I [had]been in this state I do not know. I was awakened into my senses by hearing someone moving about the cabin, apparently in their bare feet. I was under the impression at first that it was the steward. Our cabin clock was out of repair, and he would frequently come into my berth to look at my watch. I said, 'Is that you, steward? What time is it?' My suspicions were now aroused. I received no answer from the midnight marauder. I had been lying up against the skin of the vessel and when I turned to look around I noticed someone standing at the entrance to my berth. I just could discern the outlines of a man. I immediately sprang out of my berth. I had a revolver alongside of my bed. I grasped it and started in pursuit of the mysterious figure. I got on deck in a flash and looked around but I could not see anybody. So I went forward. The man that was keeping the anchor watch was on the fo'castle head. I asked him if [he] had seen anybody? He said no, he had not. I then went to the fo'castle scuttle and called down to the men. I could hear moaning and at first they did not pay any attention to my words. At last they seemed to awake and they asked what the matter was. I told them to all muster up on deck. They all obeyed. There was none of them missing. I then asked them if any of them had been out of the fo'castle during the night. They said no, that they had not. I then told them that there must have been one of them down in the cabin. They emphatically denied it.

" 'Look here.' I said to them, at the same time letting them get a good glimpse of the revolver I had in my hand. 'If I catch any of you fellows abaft of the mainmast after dark I will put a bullet into you. You have no business there. Now remember it will be at the peril of your life.' Saying this I returned to the cabin, took extra precaution to have my revolver handy, and also closed the door. And as it was now about 1 o'clock A.M., I had but a short spell of sleep before I was awoke by the steward who was now astir preparing the morning meal. I dozed away until breakfast. I got up and performed my morning ablution and set down to breakfast. Most naturally I mentioned the occurrence. The men seemed to be dumbfounded at it and never even offered a clue. I must say that I was troubled over the matter, and I tried to dismiss it from my mind as a hallucination or something of the sort. The work was carried on as usual. We never experienced any trouble with them; in fact, they seemed to be in as much ignorance in the affair as any of the after hands.

"Several weeks after this occurrence, we were lying at an anchor in Three Island Harbor in company with the bark *Roman*. The men asked the loan of the boat to go aboard of her on a visit. Shortly afterwards they returned. One of them came aft with a hat in his hand and told me that a party had fallen overboard and drowned. They had not been able to rescue him so he sunk to the bottom. I had [a] slight suspicion in the matter, so I had the boat take me aboard of the *Roman*. She was then getting ready to sail for home with a cargo of oil. I went aboard and informed Captain Swain of the matter. 'Well,' he said, 'if the man has really gone to the bottom, I will soon raise him.' Saying this he took me aft on to the quarter deck and showed me the two small six-pound cannon he had. 'They say that if you discharge a cannon near to where anybody has drowned and sunk, the concussion of the discharge

will cause the body to raise to the surface of the water.' He then
set to demonstrating the matter. First he had them swabbed out
good, as they were a bit rusty owing to the fact that they had not
been used recently. He then loaded both of them. Now they were
all set for firing. He touched both of them off at once, and the
sound was deafening and the recoil fairly shook the vessel. Two
men were ordered aft to keep a lookout that anything come to the
surface of the water. After waiting for fifteen or twenty minutes
and seeing that there was no corpse materializing, we gave it up
for a bad job. 'I rather misbelieve that the man has fallen overboard, Captain Fuller, as I have seen the experiment tested with
great success, that of firing a cannon off to make a corpse come to
the top of [the] water. It might be possible that the man has
stowed away aboard of my vessel. I will have the men take a good
look around. I would smoke the hold out, but as I am loaded it
would not be advisable.' I said, 'Of course not,' and I returned
aboard.

"The *Roman* was to sail next morning. Before leaving I was
aboard of her to send some letters home. I asked the captain if he
had made a discovery yet. No, he had not, he said. 'I have had the
men to search every available nook and corner on the ship but
have not met with any success. I assure you that if he is aboard of
this packet that he will rue the day that he ever came aboard.'
Bidding him good-bye and God speed, I left and went back to my
vessel. The *Roman* hove up her anchor and as there was a nice
breeze prevailing from the N.W., she had no difficulty in getting
out.

"To cut my narrative short, captain, I will add that [after] the
departure of the *Roman* I found out that the man had stowed away
on her. Of course, some of my ship's crew aided in abetting him. I
also came to find out that he was the midnight marauder that I had
seen standing at the head of my berth. It appears that when the
first mate went ashore after the blubber that had got adrift, he fell
into an altercation with his boat's crew. He used some very abu-
sive language on them because they would not exert themselves
more in landing the [boat] and I think he also struck one of them.

Whether it was the man 'Yorkey' he struck I do not know. Of course, I [did] not censure the mate for his conduct.

"It seems that the men held a council of war and swore dire vengeance on the mate. 'Yorkey' was to wreak their vengeance for them. He was to go below into the cabin. He was armed with a knife to stab the first mate and, if he made any resistance and awoke us, his pals were to be on the alert and come to help him put an end to all of the after hands. I really don't think that they meant to kill me in the first place. The only way that I can account for the mate not being killed is because the assassin made a mistake in the berths. At first he thought that he was at the entrance of his berth but upon seeing the chronometers, etc., in my room he found out his mistake but too late to carry out his dastardly deed. After we had been put out of the way, they were to take possession of the vessel and run her up to Madagascar and dispose of her for lucre. I doubt if they would have ever reached there as there was not a practical navigator amongst them. This fellow 'Yorkey' was an ex-convict; he even said so himself. The mate was in fault for not informing me of the occurrence ashore, and I gave him a reprimand for it. We had no more trouble with the men after we had got rid of this bad egg. I generally am on my guard against anything like this now and keep a watchful eye out for mutiny."

"Did you ever hear of this man afterwards, captain?" interrogated Captain Thornton.

"No, I did not. In fact, I never made any inquiries about his fate, although I believe if the captain of the *Roman* ever got ahold of him he did not fare extra well."

"That was a very narrow escape, I must say," said Lieutenant Ludlow. "It is very fortunate that he took your berth for the first mate's, as I am sure he would have killed you."

"Yes. They would have made quick work of all of us."

We now neared Snug Harbor. We went ashore and found a bountiful supply of good fresh water. There was a spring there and a little rivulet leading out of it. The spring was a natural one and had excellent water. We returned aboard and steamed away.

We passed by Carpenter's Harbor. Mr. Ludlow asked me its name. I told him. He then wanted to know how it derived its name from. It was some time along in 1848 or '49 that the bark *Corinthian*,⁴⁷ Captain Slate, was sailing in these parts. It seems that he was doing a flourishing business as he had one small topsail schooner, *Franklin*, collecting blubber for him from the different parts of the island. The work had increased so he asked his employers for another schooner to help convey [blubber] from one part to another and there was also a good deal of competition in the business at the time. There was material sent out from home to construct one out of. There were two ship's carpenters, my own father being one of them. They put the ship together on that spot and gave it the name of Carpenter's Harbor. The first one that they had in use here would only carry about fifty barrels of oil. The one that they made was about fifty tons burthen."

"And what became of these two schooners, captain?" asked Lieutenant Ludlow.

"The first and smallest that they built was lost on Marienne Straits; this place is the most dangerous spot there is about the island and I think myself that the individuals that had charge of her were fully exonerated. The second one met with the same fate. It is said that she was run ashore purposely. I heard the particulars of the affair. A man by the name of Williams was in charge of her. Now I come to remember *Mary* was her name. She was coming down from Table Bay to Three Island Harbor laden with blubber. They had got about half-way and the [wind] came around dead-a-head. The captain of the *Exile*⁴⁸ ordered the other vessel back. At first it was seen that he did not want to obey. His reasons — he was well known for not wanting to return to Table Bay — was because there was no rum there and he was a regular old toper — he would get into an intoxicated condition every time an opportunity would offer. Of course, there was no use in him disobeying the captain of the *Exile* as he was a superior officer, so he had to about ship and start back to Table Bay.

"He managed to get up as far as Iceberg Bay with but little difficulty. He then had to beat; it now came on night. He put one man on the lookout and gave him orders to the effect that when he

seen kelp, seaweed, to inform him of it. As you know yourself
that reefs and rocks around about this place, and wherever you see
kelp you are bound to get shallow water. But a short while afterwards the [lookout] sang out that there was kelp. The man Williams at first did not respond to [the] man [who] went up to him
and told him that there was kelp ahead. All the answer he got
from him was, 'You infernal damn fool, why do you come to
interrupt me when I am seriously meditating!' I myself think that
he must of have been in an extra meditative mood. He then told
the man to go and keep his lookout and not to come disturbing
people until he was asked about anything. The man had no sooner
taken up his position on the lookout than the vessel gave a scrape
and she ran high and dry on the beach, a total wreck, I might say.
Both of them met with bad luck. The last one that was lost was a
regular clipper as none of the vessels that come to these parts
could outsail her."

"That is a very fine site for shipbuilding. I really don't think
they could have picked out a better place."

"As for that, Mr. Swallow, I think that there are other places
quite as suitable as that for shipbuilding."

We arrived back at Three Island Harbor about 3 P.M. and we
went aboard of the *Monongahela*. The captain wanted me to remain
for dinner. I had to decline, as I wanted to go aboard and see how
things were progressing aboard of the *King*. Arrived on board and
found that they had been doing capital work — everything was
going on all right. All of the blubber we had aboard was now boilt
out and put into barrels. The mode we have of putting it into the
casks is this way. As I have already said, the blubber is minced
and put into the try pot to boil. After it has boiled into a crisp it is
taken out and put into a press and the remaining oil in it pressed.
Then the crackling is used for fuel, and excellent fuel it makes.
We have a regular fashioned pump; this is put into a cask and the
oil pumped into an oil cask and then it is stowed down. After we
had strained all of the oil down, I had the men turn to and give
[the] *King* in general a good cleaning. The boats were taken and
scalded and all of the grease and dirt scraped off of them. The
ship's deck and bulwarks were treated in like manner.

Next morning I went aboard of the *Monongahela*. She was going to take her departure for Carpenter's Harbor next day. I was invited into dinner, partook of a sumptuous meal of victuals, and after bidding the captain, first lieutenant, and the many acquaintances that I had made during her stay there good-bye, I returned aboard of the *King*. The *Monongahela* accordingly sailed next morning. And the next day H.M.S. *Volage* came in. The captain of her sent his boat off to the *King* with his compliments and wanted to know if I was going home soon and if so if I would not take some letters to the Cape of Good Hope for him. I told the officer to tell his captain that I did not know positively when I would sail as I was waiting the arrival of some vessels that were now due from Heard Island. I expected them in at any moment, and as soon as they arrived I would inform him of it. The boat then departed.

We now broke out the schooner and stowed everything properly for home. We then went over to Malloy Point for water. This was to be for the voyage home. Upon arriving there we found that the *Monongahela* and *Emma Jane* had just shortly arrived. The latter was from Heard Island. I went aboard of her and was informed that the bark *Roman* had had very poor luck and in consequence was going to remain out for another season. In fact they said that [the] whole fleet of vessels over at Heard Island had had ill luck.

I then went ashore to where the American party was stationed. I was met by Professor Kidder and invited to the house. He asked me many questions of how I was getting along and what kind of luck I had had up at Swain's Island after seal. I gave him a lengthy account of what success we had met with. He introduced me to a Mr. Holmes, chief photographer of the party, and also his two assistants. Mr. Holmes asked me if I didn't want some photos taken of the *King*. I told him yes, that he was at liberty to fire away. Getting his camera, and accompanied by the two assistants, we went down to where [the] men were procuring water. We had the big ten-barrel casks for water. These were filled up and then rolled into the surf to be towed to the ship by the boats and then hoisted by means of a tackle led on to the winch. He took several different photos of the men in different positions such as in the act of rolling the barrels into the surf and towing them aboard. Then came a picture of the *King*. He was unable to develop them on account of the atmosphere being too moist. I wish to say here, as perhaps this might come to his notice, that I never received the photos that he promised to send me. Whether he never sent them or that they were lost in transit, I have never found out. After the photos had been taken, the trio of us returned to the station. I spent a pleasant half-hour chatting with them and I then [went] aboard.

I commenced to become very anxious about the arrival of the two vessels from Heard Island, the *Roman* and *Charles Colgate*. [49] At last I got news that they were lying at anchor in Three Island Harbor. Next morning I made up my mind to sail. As we were getting underway a boat came off from the *Monongahela* with a message from Captain Thornton. He wanted me to go aboard before sailing as he had important news to impart to me. I accordingly had the boat lowered away and went aboard of the *Monongahela*. I was met by Lieutenant Ludlow. He informed me that there was a boat's crew from the *Emma Jane* aboard. They had arrived through the night in one of *Jane*'s boats, so they must have taken "French leave." They had entered a complaint against Captain Rogers. I then was conducted into the presence of Captain Thornton. I asked him what the trouble was with the men of the *Jane*.

"They complain that they are detained down here in these parts against their wills and contrary to the agreement on the ship's articles. They state that they only shipped for one year, and the specified time has expired. The vessel is to remain here for another season, but it is contrary to their wishes to remain. Now you know yourself that the law of the United States won't allow men to be detained in any part if it is contrary to what they shipped for. I thought I would get your advice on the matter as you are much better posted on matters as regards shipping to these parts."

"Captain Thornton, I am not acquainted with the agreements of these men and I do not know how the articles of *Roman* are

drafted as they might differ in certain respects to the ones that I
have come in contact with. Suffice to say, that the men that I
shipped and the ones I have shipped to these parts don't ship for a
specified time but until the vessel gets ready to return and it is at
the captain's option to let his men go. In this case I think it would
not be advisable to part with his men as he won't be able to procure any. The whole long and short of it, captain, [is that] I think
that the men have misrepresented the case in question to you. I
really don't think that it is any fault of Captain Rogers as he had
nothing to do with the shipping of the men and of course he has to
adhere to the ship's articles. Captain Swain had the crew shipped;
Captain Rogers is his successor. If there is any fault to be found it
is in the owners, as I think Captain Rogers can be freely exonerated."

"I did not come down here," said Captain Thornton, "to interfere with shipping matters, and as you are going over to Three Island Harbor, I wish you would oblige me by telling Captain Rogers to come over here an to bring his ship's papers with him, as I
want to see into the matter of these six men. I will be obliged to
arbitrate the case and see it adjusted in the way that it should be."

I told him that I would. After a few words more on a variety of
topics, I bid him good day and went aboard of the *King*. We were
but a short time getting underway. There was a good breeze prevailing and we stood off with all sail on for Three Island Harbor.
The distance being but short we got there in a very short time.
Upon getting everything fixed aboard I had the boat lowered
away, and I went aboard of the *Roman* to deliver Captain
Thornton's message to Captain Rogers.

Upon getting alongside I was met by Captain Rogers and invited down into the cabin. I informed [him] of the complaint the
men had made against him. He told me that the men had misrepresented the matter. "At any rate," he said, "I will have an
interview with Captain Thornton."

It was now going on towards evening. The sun was but a few
hours from the horizon. I got up to leave and Captain Rogers prevailed upon me to stay for tea, so I finally accepted. He seemed
to be very much agitated over the matter of the men absconding

and he wanted me to advise him on different points. He was a peculiar character. He was a man of about fifty years of age, although a casual observer would not take him to be over forty-three or so.

To characterize him would be a hard matter. There is no doubt but what he was inclined to be eccentric. He seemed to delight in keeping up a foppish or rather extreme stylist appearance. This little fact will prove it. When we went aboard of the *Monongahela* at Malloy Point, he was decked out in his best Sunday go-to-meeting broadcloths, boiled shirt on, cutthroat collar, toothpick footwear, watch and fob, hard hat, and, the most ridiculous of all, he had his "fish" waxed and his hair combed in the center, a regular "laddie oh." It was absurd for him to have gone to all of this trouble. He should have saved his cosmetics, etc., for some grand ball or better for his own funeral. I hope the reader does not misconstrue my words; in fact, style — boiled shirts, etc. — I think are things that are indispensible in the life of a man when [on] shore, to be used on certain occasions. Even whalemen should make use of them. But the idea of trying to put on extra style in a remote and nonpopulated placed like Desolation is absurd to the extreme. A person to have seen him would have imagined that he had decked himself out in all of his finery to attend church or better still for an exhibit on the manly art of self-defense.

It was just sunset when we were invited to sit to tea. I and the captain were the sole occupants of the board. He still kept discussing the matter of the runaway seamen. After we had finished he invited me into the after cabin, his own private apartment. I had just got below and seated myself and he stepped into an after berth and, hello, he ushered a being of the feminine gender into my presence and introduced her to me as his wife. I was dumbfounded at this; now this was a sequel to his reason for being so particular in dress.

"Why, Captain Rogers, I was not aware that your better half accompanied you." He answered me in a *sang-froid* manner. I then addressed his wife. Her health was excellent, she said. "And how were you impressed with Heard Island, madame?" I asked.

"Really, Captain Fuller, we had a delightful stay there; it [is] such a romantic, I might say remote place. I evinced an awful lot of interest in it while there."

"Yes," thought I, "you would alter your opinion of the place if you were obliged to undergo what we men folks do." "Would you fancy ever returning to these parts, madame?" I ventured to ask her.

"As you know I have all of [my] affection centered in my darling here. Wherever lovey goes, I am satisfied to go with him."

A long string of thoughts then commenced to run through my mind, family quarrels, henpecked husbands, and worst of all divorce. I thought it very curious for Captain Rogers to bring his wife to these parts. At any rate it was a very poor place for spending a honeymoon or a bad trip for a bridal tour. None of your fine parks or balmy twilights in this part of the globe for lovesick people to coo over in. I myself did not think much of this gumsucking lovesick pop-i-cock, as Captain Sisson termed it after me telling him about [it].

We had quite a lengthy conversation on topics of no interest. She would insist upon talking family matters to me. She was a comely looking dame of about four and twenty, of about medium size. She claimed to be a school mistress, and hailed from some western state. I don't wish to say anything against Captain Rogers; nothing personal, but I don't think it was much of a catch for her when she wed him, he being a man advanced in years. I suppose she got fascinated with the notion of traveling over old Neptune's Domains, in a windjamming whaler, and entered the holy bonds of matrimony. Poor woman and may the Lord have mercy on her soul. I heard afterwards that she died on the Island of Mauritius, Isle of France, in giving birth to a child. Whether the child survived her or no I do not know.[50]

Not feeling very deeply interested with the couple, I took my leave. Before leaving, Captain Rogers asked me if I would not accompany him to the *Monongahela*. I answered by telling him, if it proved convenient to me. I was then going aboard of the *Colgate* to see Captain Sisson in regard to getting my oil home by him. I had some six hundred odd barrels that I could not possibly accommo-

date. I got into my boat and went aboard of the *Colgate*. Captain
Sisson agreed to take my oil. It was then that I related the love-
cooing business that I witnessed aboard of the *Roman* between
Captain Rogers and his better half. It seemed to disgust him out-
right. "And he wants you to go with him to see Captain
Thornton. It's a wonder that the old reprobate has not learnt to
fight his own battles."

"It is not for that purpose that I am going to accompany him,
but more to oblige Captain Thornton of the *Monongahela* and as I
want to see some parties on different matters. I might as well
humor the old fellow by going with him." Bidding him good evening, I was pulled off in the boat. I had the man pull close on to
the *Jane*. I hailed the captain, and told him all right, that I would
go around with him on the morrow. I would be aboard at about 6
o'clock A.M. "Many thanks, captain," he said. I could see that he
was very anxious about me going with him. I arrived aboard. I related the little coincidence to the after hands, and I must say that
they had a good hearty laugh over it.

Next morning found me aboard of the *Jane*. Everyone was active there. We got underway and ran off before a good stiff northerly breeze, arriving at Malloy Point at about 9 o'clock. We went
aboard of the *Monongahela*. Of course, Mr. Rogers was decked out
in all of his finery. I must say that he was quite a contrast to the
ungainly looking surroundings and he created quite a sensation. I
heard some of the men passing different remarks about him. One
of them asked a fellow shipmate if that 'ere bloke thought they had
church service or a wedding aboard that day. [Another] said,
"That is Reverend 'thinks-he's-somebody' come to officiate at Di-
vine Service today." I escorted him and Captain Bailey — who
had just arrived — into the presence of Captain Thornton. I introduced them and then excused myself.

As I came on deck they had just got through completing arrangements for target practice and had just fired a volley or two. I
had been there but a short while before I was joined by Captain
Rogers, who looked very jubilant. He had the appearance more of
[a] tinhorn gambler or ribbon clerk than that of a skipper of a
whaler. I asked him what success he had met with. He said that

matters were fixed and that the men would be obliged to return aboard. He appeared to be highly elated at this, I must say. He asked me to go forward with him and witness closer the proceedings of the target practicing. No, I told him that it would be taking undue liberty. It happened that Lieutenant Ludlow was just passing and he overheard my remark. He came up to me and said, "Go to any part of the ship you wish, captain; you are considered a privileged character aboard of her." I thanked him very much and we went forward and seen the men go through a regular role of man-o'-war tactics. They did some very good shooting and seemed to understand the management of the guns perfectly.

When they got through practicing we were joined by Captain Thornton, who asked me when I was going to leave the island. I told him next day if everything worked satisfactorily. I then asked him when he thought he would get away. "It's hard to tell, captain, although I think I will sail in two or three days, and as you are going to get away tomorrow I will send a boat around to the English Station to see if they have any mail that they want you to take to Cape Town." I told him that there was no necessity of [him] going around to the English Station, as I was going around to Three Island Harbor myself.

"Well, you will please ask the captain if he wants to see [us] on anything special to let me know and I will go up there at 5 P.M." I told him that we would do as he had bid.

Captain Rogers and myself went aboard of the *Jane* and she stood away for Three Island Harbor, arriving there at dark. During the [succeeding] five days we were busy getting our oil aboard of the *Colgate*. [51] We managed to get all we had on shore, about six hundred barrels. While we were getting the oil [on] board, part of the crew were engaged in making final preparations aboard of the *King*. We were all ready for sailing when the *Monongahela* arrived in the bay. I went on board to see Captain Thornton and he said, "Well, Captain Fuller, are you all ready for sea? I see that you have everything in readyance to leave, captain. When do you expect to get off?"

"I will be getting off tomorrow, Captain Thornton, if everything proves favorable." Next morning the weather turned out to be very fine; went aboard of the *Jane* to bid Captain Roberts *adieu*. I found that he had the men that had taken "French leave" in irons, and he had taken all of the sails out of one of the lockers and had turned it into a dungeon. The men were atoning for their crime in this place. I bid him and his wife good-bye. I then went aboard of the *Colgate* to take leave of Captain Sisson. I found him in good spirits. He seemed to be in great elation about getting home once more as he said. He was to go via the Cape of Good Hope. I then went aboard of the *Monongahela* and Captain Thornton asked me if I was going to get under way soon. I told him that I was going to leave immediately.

"I am going to have a race with you. I think that the *King* is a very good match for the *Monongahela*."

"Yes," said Lieutenant Ludlow, "we will give you a race. I am afraid only that you will be victorious. From what I have seen of the *King*, she is a clipper, a very good sailor."

"But you must remember, gentlemen, that you have an advantage of me; you can use steam whereas I am at the mercy of the winds."

"We won't use any steam, I assure you. We will not take that advantage of you — no steam after getting clear of the land — but of course [if] any unlooked for accident occurs we will be obliged to make use of it."

"Well I will bet on the *King*," said one of the young lieutenants.

"I assure you, gentlemen, that if the *Monongahela* does not use steam that I will give you a tight race to the Cape. And I have an idea that I will beat you."

"Bravo, Captain Fuller," said Lieutenant Ludlow, "I think you will myself."

After bidding them good-bye, I went aboard of the *King*. I found the men from the bark *Roman*, three forehands, two officers, and one boatsteerer, had come aboard. They were to go as passengers. Everything being in readyance for sailing, I had the anchor hove up, and we started on our way down Royal Sound. An hour afterwards the *Monongahela* came steaming down. We dipped colors to each other, and she steamed on out to sea. I must

now say that we were making a very good commencement to gain
the honors of the race. We had no sooner got outside of the foreland than the wind commenced to blow heavy from the N.E. and
to make it more disagreeable it commenced to rain in torrents and
became very foggy and thick. I could see that it was foolhardiness
for me to try to get clear of the land so I ran down to Shoal Water
Bay. I had the anchor let go here and laid in for the night. We
were a disgusted lot of men. Of course there was no fear of the
Monongahela not being able to weather it, as she was using her
steam to get clear of the land in the first place. This was a great
advantage he had over us. The prospects of us beating the *Monongahela* were very gloomy, as we had not the slightest idea when the
gale would subdue.[52]

    [In] the morning just before dawn, the wind lulled and
chopped around to the S.W. At daybreak we hove up anchor and
got underway again. We had been underway only for about an
hour or so, when one of the men came running aft in [a] hurry. He
seemed to be very much excited. When he got to the brake of the
quarter deck, he could not hold it in any longer but sang out that
the vessel was leaking terribly forward. I immediately summoned
the first mate and told him to go and investigate the matter. He
went forward and returned shortly and informed me that there
was a steady volume of water coming in forward as large as a
man's arm. I gave orders for the men to convey the spare anchor
we had and all of the chain as far aft as possible. This was to bring
her head out of the water. I then went forward and examined her
myself. I found that the leak was around the hawse pipes. Seeing
that it was an impossibility to remedy it out at sea, I had her put
about and stood off for Three Island [Harbor], this being about
the handiest place. It was not anything serious as it could be fixed
if we could manage to get into smooth water where she would not
heave so much.

    We got back to Three Island Harbor at dark much to our own
disgust, but we had to reconcile ourselves to cruel fate. Now this
was the second time that we had been obliged to take refuge in
harbor. First the merciless ill winds and now the vessel had commenced to leak when we had just got started well on our journey.

It was pitch dark that night, so we could not do anything with the leak. We were obliged to leave it until morning.

Next morning arrived. We proceeded to investigate the leak. After a little examination we found that one of the hawse pipes was leaking very much. It had started out — got sprang — in some manner. We drive it back in, and had it caulked tight and put battens around both of them. We arranged everything aboard again and it was high noon when we got underway again. There was a fair breeze prevailing from southard and westard so we experienced no difficulty in getting out clear of the land. Arriving outside, we tried the pumps and found that she was good and tight, so we let her have all sail that she could carry, and we went fairly skimming over the water.

Nothing of any interest transpired during the passage up to the Cape of Good Hope until we were off of Cape Agulhas — the most southern extremity of Africa — [where] we had two or three heavy blows. Luckily we were not obliged to lay to in any of them. We fell in with the schooner *Charles Colgate* off of Cape Agulhas. The Cape Agulhas be in 34° 49′ Lat., 20° Lon. and the Cape of Good Hope 34° Lat., 22° Lon. The former one is one of the most noted places for blows there is in the globe. The sea that prevails around this point is very dangerous. It is not so large as the Cape Horn sea, but very choppy and small and sharp, which makes it more perilous to be in in a blow. The blows they have there are called Cape "snoozers."

We were right abreast with the *Colgate*. We were both on the same tack, she being to windward of us and about an eighth of a mile ahead of us. We were going to race it into Cape Town. We sailed along together, neither of us gaining on each other. We got up to the entrance of Table Bay and now came the break for the anchorage. As I have said, the *Colgate* was to the windward of us.[53] A heavy S. Eastern was prevailing, so we were obliged to beat in. When we got between Green Point and Robin Island, the wind lulled down, but only momentarily, for it breezed up again, and so sudden that it sent both of us on our beam ends. We were not obliged to take in any canvas, but both of us sailed right in on our starboard tacks and dropped anchor at the same time. Large

numbers of people had congregated on the beach and on the pier
that we came to anchor at. This pier is situated right at the foot of
Adderley Street.

As we were coming in we found that the U.S. *Monongahela* had
already arrived, so our race was lost. We had just got the anchor
down when a boat came off from the *Monongahela* in charge of
midshipman Marryman. He said that Lieutenant Ludlow sent his
compliments to me and wanted to oblige him by letting him see
my chart as he wanted to see the route I had come from Desolation. I let him have it and returned his compliments, and told him
to tell Lieutenant Ludlow that I would try and be aboard to see
him as soon as I fixed matters ashore. Luckily the harbor master or
captain of the port, as he is known in English places, was not long
in coming aboard. Usually other times he would keep [me] await-
ing until I got disgusted with the place.

It is a rather curious matter how these harbor masters treat
American vessels or "Yanks" as they term them. It seems to be the
same in all parts of British possessions. Now, for an example, if an
American vessel enters a British seaport in company with any
other foreign vessels, they might be Scandinavian, Dutch, Italian,
or Spanish, he will invariably come aboard of them first and leave
Jonathan for the last. He will even put himself out of the way to
carry out his spite. I find that the Americans are retaliating; they
do the same thing to Johnnie Bull. But this time the harbor master
was very prompt in coming aboard. He found everything all
right, so I went ashore. I went up to the American consul's for my
mail as I had had it directed to his care, received several letters,
[and] got good news from home. I visited a few friends, took tea
with one of them, and got back aboard at 6 o'clock P.M. The rest of
the evening until bedtime was occupied in answering letters.

Next morning I was aboard of the *Monongahela* at about 10
o'clock. I was met by Lieutenant Ludlow. He seemed to be glad to
see me, and he was in a very talkative mood. He invited me into
the stateroom. We both seated ourselves, and he said, "I suppose
you have already heard of the accident that befell Captain
Thornton?" I told him that I had not heard any of the particulars
of the case. He then told me how Captain Thornton had give a

slip when the vessel was heaving and injured his spine. In fact he said that it was thought that the injuries he had received were of a serious kind. They were off of Cape Agulhas. They were having a little blow and the vessel was heaving considerable, and the Captain walking about in his cabin. The vessel gave a sudden lurch. He fell up against an article of furniture and he hurt himself very much. Seeing that he could not possibly recover during the stay in Cape [Town], his injuries being of a serious nature, he was sent home on one of the Royal Marine ships via England.[54] I must say that I felt sorry for Captain Thornton as he was a very nice man, a proper gentleman in every respect.

"I was looking over your chart last night, captain, and I find that you only have twenty-eight days? As you have already found out, I suppose, we were twenty-six days. And I am sure that you would have beat us if we had not made use of our steam." I then told him how I had made it twenty-eight days on account of having to run into Three Island Harbor to fix the vessel when she sprang a leak on us. They experienced but little foul weather on their run up. He told me, as they were very eager to get up to the Cape, they made use of steam.

"I suppose you have seen the morning papers, captain? They give you and the *Colgate* quite a puff." He then picked up a copy of the *Morning Argus* published in Cape Town and read quite a lengthy article to me. It was headed thus: "A race into Table Bay by two American Clipper Schooners, the *Roswell King*, Capt. Joseph J. Fuller, and the *Chas. Colgate*, Capt. Sisson, both vessels flying the stars and stripes and hailing from [New London]. Beating and tacking in against a heavy S. Easter." The above was the heading of the article, then the article dwelt upon different maneuvers gone through while beating in. Lastly it said that it was a rare sight to witness and was watched with interest by hundreds of people.

After he had read this, he said, "I was watching the proceedings myself and when you got becalmed between Robin Is. and Green Point I was going to send the ship's steam launch after you to tow you in, but I could see you were having a race and I wanted to see you beat the other fellow. It happened that the harbor master was aboard at the time, and he seemed to watch the race with interest. We seen you when you were making in first. I knew that it was the *King* that was to the leeward of the *Colgate*. The *Colgate* held this advantage over you, and besides she was about one-eighth of a mile ahead of you and she had more sea room to tack in. The harbor master said, 'I am afraid that your favorite, that fellow to leeward, is going to be bested.' When the *Colgate* tacked and stood off towards you and when you got in such close quarters, he said, 'Great guns, he will be into her.' You were now pretty well over towards the Blueberg shore, and both of your vessels' channels were buried deep in the water, and when the *Colgate* got close into Blueberg shore, she tacked but you kept on straight, much to the astonishment of the lookers on, and I dare say to the captain of the *Colgate*. Now the harbor master grew very enthusiastic over the race. He almost yelled out at the height of his voice: 'Great guns' — this seemed to be a favorite by word of his — 'he will run afoul of her, sure. Why does he not keep off and run around her stern?' As he said this the *King* flew up in the wind, and I must say astonished us all not a little. When your jib boom was over the *Colgate's* taffrail, and when the harbor master said that you would be into her, I told him not to be quite so quick in coming to [a] conclusion. As a man that cruised in [those] parts that you had just arrived from, [I] understood what [you were] about. Clear from the time that you come in sight of the bay and until you anchored, you were both watched with intense interest. In fact, everyone evinced a good deal of interest in the affair. And I will say that you did honor to your colors."

We partook of some wine during his recital of the incidents of the race. He told me that Captain Thornton had left several parcels with him and he wanted to know if I would not take them with me. I told him yes, to send them aboard, and I would see that they reached their destination. I then got up to bid him good-bye. He returned my chart to me saying that it was a very creditable run for me from Desolation. I bid him good-bye and told him I would see him before leaving.

"By all means, captain, you must see me before you leave. We are to have an audience with his excellency the governor tomor-

row. I suppose we will have a lot of John Bull red tape and cour-
tesies to undergo. And I learn that we are to return to our former
station, Montevideo, from here. Good-bye," he concluded, and I
went ashore.

I started up town and as I was making my way to Plein Street,
by the way, [the] principal commercial thoroughfare in the place,
I was accosted by a Mr. Norton. He was then the editor of the
Cape Argus. I had an introduction to him during a previous visit to
this place. He prevailed upon me to go and take dinner with him,
so I accompanied him to his private residence. He entertained
very hospitably. He seemed to be very deeply interested in matters pertaining to the fishing and oil trade. I separated [from] his
company and thought I would take a stroll up to the museum. I
had heard that it was quite a place so I though I would not miss
seeing it. They also have a public library in connection with it. I
walked up Adderley Street. [I] passed by the Parliament House, a
very fine and commodious looking ediface. I then got into Garden
Street.

After a pleasant walk through a boulevard, as the French call
it, or grove, in English, I arrived at the museum building. It was a
very substantially constructed building from its appearance and
very nicely situated. It fronted the botanical garden. Trees and
shrubs and flowers were growing profusely about the place and
the air was laden with their perfume. There is a stair [of] very
broad steps leading up to the entrance. I ascended this and entered. The first specimen that was presented to view was a familiar one. They had an enormous big sea elephant stuffed and looking as big as life. Of course, this was not much of a hors d'oeuvre
to me. They had a good many sealike creatures fixed up in the
same way. I had not entered the museum proper yet, this only
being a hall that led into it. I was looking into one large glass case
that contained several different kinds of African bucks and a
specimen of petrified wood from the Petrified Forest in [the]
U.S.A. Near my gaze, alongside of it, was a small idol hewn out
of slate from the Zimbabwe ruins in South Africa. The ruins are
supposed to be the remains of King Solomon's Mines.

I now entered the museum proper. It is quite a large affair and a credit to South Africa, they say. Herein [are] exhibited all of the present existing animals of South [Africa] and many of the antediluvian fossiled kind. It would be an impossibility for me to give a description of even [some] of the most interesting ones. All of the natural history studies are represented here, principally geology, zoology, ornithology, etymology, botany, piscatology and also the various other studies. It was quite a treat to go through the place. The whole thing makes an exhibit equal to many of our museums at home.

After satisfying myself with the museum I went to the library. This is a public one and any one is admitted gratis. They have quite a number of very fine paintings in it, and a very large and fine assortment of books. They receive newspapers and magazines. In fact they have periodicals from all parts of the globe. I picked up a copy of the *Detroit Free Press* and had quite an interesting moment to myself. The attendance at the library was a good many. Hearing the hour of 3:30 P.M. struck, I had entered here for a good three hours without noticing it any, having been so interestingly absorbed in what the museum and library offered to view. I left the [library] and went out on to the front steps, these being sheltered by a very high portico. I could get an excellent view of the botanical gardens. It is quite a garden, covering some six or seven acres of choice ground. Not having much of a taste for the study of botany, I did not take a stroll into it, although I think it is a very interesting place and quite worth while visiting.

Having ample time and feeling in a mood for taking a walk, I left the museum building and walked at a brisk pace up the Garden Street. Here were a good many pedestrians walking out more for pleasure, so I was not entirely left to myself, although at times the walk was quite deserted. Boers — native farmers — I often met.[55] They were just in from the adjoining country and were taking in the sights. Now and again I would meet a Hottentot, Kaffir, Malay, or perhaps a Zulu. Cape Town is a very interesting place for different sorts of people. Here the European — they were represented from their various countries, Africander — native white African — Hottentot, Kaffir, Zulu and, I dare say,

American mingle together making a regular conbluberation of lands, religion, and sexes.

The white element, of course, are invariably Christian-thinking people and also the blacks such as the Zulu, Kaffir, Hottentot, etc. But the Malay is a firm believer in the call of Mahomet. The Koran, like our bible, is what their religion is founded on. I have heard of Europeans joining them, of course more for a policy than for sincere belief. The general run of dress here is much the same as what it is in the States. Even the black after a little civilizing takes to pants and shirt and hat. But the Malay, well they seem to be a curious sort of being; they still seem to hold to their mode of dressing to a certain extent. They wear the bell-top fezzes and have the wooden shoes with fore and aft heels. The priests dress very gaudy and also the women. The former wear a long frock gown made out of velvet and silk stuff and variegated colors. The women delight in gaudy colored silk kerchiefs for headgear and very stiffly starched and commodious dresses, which makes them look twice their natural size. They dispense with the wooden shoes on certain occasions and adopt the neater and more tasty looking footwear of other civilized nations. They seem to have a weakness for fancy patent leather shoes and like their sisters, the Chinese, are admirers of very small pedal extremities. In matters of religion they are very sincere, and the laws laid down by the Koran are very stringent. The women, I learn, are very virtuous and make excellent domestics, while the men are among the best artisans and tradesmen of the place.

I now arrived to the terminus of the walk, and a splendid view of Table Mountain was offered. The summit of it was enveloped in clouds. They were rolling and turning about, which made it a sight never to be forgotten, and, to add more grandeur to the scene, the whole surrounding range was covered with a coat of green vegetation. It had been raining on the mountain a bit and the sun was reflecting down through the clouds. Cataracts of water could be seen flowing down ascents abrupt enough to be called precipices and then the reflection of the sun on the silver leaves of the trees made it a grand spectacle to view.

Among these mountains can be seen vast swarms of monkeys of the baboon kind, also the roe buck, spring[buck], and zebra are to be seen at intervals and once in a great while a lion or giraffe. Of course, other animals of the feline and canine tribe are to be met with here such as the lynx, African tiger, wild cat, jackal, wild cow, fox, etc. These seem to be very abundant, and the government offers a bounty for the scalp of each one that is destroyed. Knowing what the overhanging clouds on the summit of Table Mountain meant, as it is a well known thing when clouds overhang this mountain it is a sure indication of a brewing storm, I journeyed back to the vessel. I arrived there after a walk of some thirty minutes. I took a shore boat and was rowed aboard. Found everything all right.

Next day I gave the men a run ashore. They acted all right ashore, none of them absconding, and all returned looking about the same for the different enjoyments they had indulged in. Next day it was a memorial one, American — Washington's birthday — I could see the *Monongahela* in the distance. She had extra ensigns flying and her code of signals out. I took advantage of a S. Easterly breeze that was prevailing, hove up anchor, and got under way. I made it a point to run down by the *Monongahela* and as we were passing her stern Lieutenant Ludlow was on the quarter deck. He was the first to hail me with, "Good-bye, Captain Fuller, and a speedy and pleasant passage home for you." I wished him the same and told him to report me when he arrived home. He would, he said. The officers and men of the vessel waved their hands at us and we passed on out of speaking distance. We were a merry set aboard of the *King* as we were homeward bound, and the men had quite a little sum coming to them. We sailed along under a good stiff breeze, and by evening we were good clear of the land. We were to go via St. Helena. Luckily we got the S. East trade winds, which proved very favorable. Our run from the Cape there was some nine days. A very fair passage.

We did not come to an anchor here but laid off and on. I went ashore. I met my brother-in-law, a Mr. Adams. After arranging things with the captain of the port, I went up to my brother-in-law's residence. They were glad to see me and asked me many questions about home and the folks. I did not take in any of the sights this trip, that is, such as Napoleon Bonaparte's residence and tomb or the country house of the Zulu chiefs. No doubt but what the reader has read and heard a good deal about the former place so I won't make more than an allusion to it here. The house in which this noted warrior spent his last days is some three or four miles from Saint James and the tomb is some two and a half miles distance from the house. They are in good preservation and are looked after by a retired French military sergeant, who resides near the place with his family.

The place where the Zulu Chiefs are confined is at . . .[56] and is some . . . miles from Saint James. This chief is here accompanied by . . . of his best men. They also have their women with them. He receives an allowance of £800 — $4,000 a year. He is lodged in a very good house [of] two stories, fitted out with all the modern household improvements. They are exiled for creating a revolt among the natives in S. Africa. They are here for different terms. There is no doubt but what they are contented with their lot, at any rate we are left to think so. The king, I don't recollect his name, is a swarthy looking negro of about thirty-two years of age and of an excellent physique.

Other notable sights about the island are Ruhert's Valley [and] Diana's Peak, the largest mountain on the island. Also a place worthy of note is the old English battery facing the sea from Saint James's side. The ladder, or Jacob's ladder as it is known here, extends from the bottom of a good sized mountain to the summit. It leads to the soldiers' garrison. It has some 900 steps. The majority of the natives are of English and negro extraction and are known as "yam stacks." I don't know the meaning of this term and have not been able to find out what it implies so the reader will have [to] remain in ignorance.

The primary occupation is farming, the soil being of a very fertile nature. Vegetation grows very luxuriantly. Vegetables and fruit are grown to a large extent here. Both tropical and temperate fruits are produced. They are just after introducing the coffee tree and they seem to be meeting with great success. The people are very hospitable and courteous to strangers and seem to be al-

together very industrious. The island is presided over by a governor appointed by her majesty, the queen of England. In late years
the place has become of less importance to navigation. Like other
places lying in the S.W. Atlantic, they have been seriously affected by the Suez Canal. Formerly before this was in existence,
ships bound for the East Indies and Antipodian places from England or the Continent were obliged to run around the Cape of
Good Hope. They would either call in at Saint Helena or Cape
Town. There are only a few sailing ships that call in here and once
in a great while a steamer that is a slow mailer, as there is a regular
line from London to Cape Town and they run into Saint Helena
bimonthly.

It was now getting on towards evening, so I gave directions for
everything that I had purchased ashore such as potatoes, cabbage,
etc. I received a letter from my wife and found out that all was
well at home. At 4 P.M. I got aboard. Having a splendid breeze in
our favor, we set all sail and went on homeward. We had extra
good weather during the [succeeding] days; we sighted the Island
of Ascension — British — [it] is about [700] miles to the [north-
west] of Saint Helena. It is of less importance in certain respects
than the latter. They have troops stationed there and use it as a
man-o'-war station. It is inclined to be barren, and is about one-
sixth smaller than Saint Helena. We had a very fair run from
Saint Helena to the equator, some eleven days, and were very
lucky in escaping the doldrums. We were becalmed for a couple of
days but it began to breeze up and we found ourselves in the
North East trades. We held these winds until we got up around
the Island of Bermuda, some seven hundred miles from New
London.

One morning it commenced to breeze up from the N.W. We
had been having a calm, only a bit of air now and again. It came
on to blow a living gale at forenoon. So I had her hove to. There
was a terrific big sea a'running and the vessel would dive down
into the sea and large volumes of water were continually coming
over the ship fore and aft. On the morning of the third day, the
wind hauled around to the S.S.W., so I kept off and tried scud-
ding before it under short sail. As I have said, there was a tremen-

dous big sea a'running and I might say that at times the whole vessel would be submerged in water. At last one very large and extraordinary one came along. It not only covered the decks with water but broke in our foresail and carried it away. We were obliged to take it in and replace it with the storm trysail. I then had them take the jib and bind it on to the mainmast for a main staysail. We then hove to again and as soon as everything was made snug we set to work to repair the foresail. We managed to finish it before night. Through the night and next day the wind slowly moderated and the sea commenced to go down. This being the case I kept off and run before it.

We had been running but a short time, perhaps two hours, when one of the [men] came aft looking very much excited and said the water was running into the fo'castle. I sent the first mate to investigate the leak. He hurriedly went below into the fo'castle. He had been gone but three or four minutes when he returned and said that the water was completely deluging the fo'castle. I asked him if he managed to locate the leak. He said yes, that the water was coming through one of the seams about four feet long and several inches wide. I immediately gave orders for them to man the pumps. They pumped away for a good bit but they could not even manage to get a "rolling suck." I seen that it was an impossibility to free her of water on this tack. So I had her put about on the southern tack. I gave orders for the pumps to be manned again. After a good deal of hard pumping, we managed to free her. I then went forward myself and after a good deal [of] trouble located the leak. I found that the planksheer had been hauled up, the oakum had been forced out by the water and consequently had sprung a leak! It was impossible for us to remedy this at the present, so I had her hove to. I had the men continually at the pumps.

Towards evening the sea commenced to go down and by 5 o'clock, I though it smooth enough for us to try and fix the leak. I asked Mr. Joseph if he thought he could caulk the leak. If he could I would have some of the planks cut away around the fore chains so as to admit his body and then he could get full play and get up against the planksheer. He said that he thought he could fix the leak, so he went to work.

First the planks were cut away from the bulwarks. I thought that it must come up with the forward shroud so I had the foresail took in. He managed to get at the leak, but it was a difficult thing to get the oakum to stick and stay in the seam, so I had them take some half-inch rope and drove it through some tarred canvas and then I had it tacked over the seam securely. This had the desired effect, for it quit a'washing. To make it more secure I had some battens — made out of some old boat boards — nailed over all. It was a hard job as Mr. Joseph was willing to testify to. The vessel would root her nose down into the sea. Of course, she would be bound to ship an enormous lot of water and it would cover him completely up. Then she would come up in the wind, a big sea would come rolling along over him and almost shatter every bone in his body.

He persevered until the work was complete. He was a sorriful looking spectacle when he came up on deck. He looked like a drowned-out rat. After I found that the leak was remedied, I had them set the foresail again and then we went about on the northern tack, had the pumps tried, and found she was as tight as usual. I did not dare to attempt to carry all sail as I did not know at what moment something else might give away.

We managed to fairly get over the water with the sail we had on and in three days we got into the latitude of the Gulf Stream. On the fifth day the wind came around N.W., dead in our teeth, so I had her hove to. Next morning we kept off. About 5 P.M. we came in view of some coasters. They were apparently standing offshore in the wind. We stood along so as we would come within speaking distance of them. We got within speaking distance of them, so I hailed the closest one. "What way does Montauk Point bear from here?" I asked him.

"Darn you, you are not heading for Montauk Point, but Cape Cod, bearing north."

I could see that the man was intoxicated, so I [replied], "Thank you. How is [the] rum?"

I laid along close enough to another one and got the desired information from him, so I stood away for Block Island, which he said bore north from me. We had a very nice breeze from the southern and westard and in one hour sighted land, and by night-fall we were hugging it pretty close. It came very thick and commenced to rain, but I managed to get my bearings off Little Gull Island light. I had not determined whether to try and get in or to lay to for the night, when Mr. Usher came up to me and said, "Keep on, captain, and I assure you that I will take you in safe. I am acquainted hereabouts and you need not have any fears of me running you afoul of anything."

"Go ahead then, Mr. Usher," I replied to him, so we stood off for the sound. We had been sailing along at a brisk space for about three hours, when we seen a light on our starboard bow. I asked Mr. Usher what light that was. He replied that it was Race Rock, and added that we would see New London bearing north. We kept standing along and in about one hour's time, we seen [the light] north of us. We headed due for it, passing close to the light that he said was Race Rock. I had give orders to Mr. Francis to keep the lead a'going. Of a sudden the light that we had been heading for vanished. I sung out to Mr. Francis, how much water had he? And he replied. "Six fathoms."

We kept standing along for five minutes more. The light that we had lost now reappeared. It happened that some object had got between us and the light so it had been obscured from us. We now gave her plenty of headway, and we had the helm hard down, ready to tack, when Mr. Francis sang out, "Put your helm hard down, six feet of water!" The helm was immediately put hard down and she scraped her side up against the reef that he had sounded on. My God, but his was a close shave. A couple of feet more and we would have been dashed up against the reef. Of course this put me on my guard for any further accidents. I told Mr. Usher that I would run no more risks but should anchor there for the night. We took soundings and found that there was six fathoms of water and the anchor was accordingly let go. I must say that the suspense was awful. In fact I was so eager to get in that I could not satisfy myself by going to bed, but employed my time in making arrangements to go ashore by packing up. It was a case now of yet so near and just the reverse. Everybody else seemed to be as eager as myself to get on *terra firma.*

I managed to find enough packing up to do until dawn. I then went up on deck. At first I couldn't get my bearings. The place seemed familiar to me as it grew lighter and I could get a better view of the place. At last it dawned upon me where I was, not as I thought in the Race but in Fisher's Island Sound. I must say that this was a narrow escape from being grounded. Of course, it was Mr. Usher's fault, but as we had no harm befall us I did not feel like reprimanding him for it. I don't believe he had a very good idea of this part of the coast and besides I suppose he was tempted to make the attempt to pilot us, as he would have received some recompense for it in the shape of cigars, tobacco, or something of the sort.

Shortly after breakfast the pilot arrived aboard. He also seemed to be astonished at where we were. He asked "How come you in by Water Hill Light?" I told him how it had occurred; we had been deceived by a light and Mr. Usher getting the wrong bearings.

"Captain," he said, "you were very, yes, I will say extra lucky, in getting through the reef without losing your schooner. It is a difficult matter to get through in the day time, not much less at night."

We arrived at the terminus of the New London Harbor [and] found a steam tug waiting for us. I had all sail taken in, hove the warp, and was carried up to the wharf. After adjusting matters there that night I took the train for home. Arrived in Danvers that night and was welcomed by my wife and family, who I must say from appearance and joy had felt very anxious about me.

So ends the details of my "Four Voyages." I am still following the sea. If this work meets with a general acceptance by the general public I have in course of completion one that I think will prove of more interest than this one, as it treats on the perils that I underwent in a shipwreck and deprivation. We — myself and twenty-two men — were cast away on a desolate island, and after undergoing untold suffering, etc., and being there for some eleven months, we were rescued, all surviving but four.

PART II

The Wreck of the
Pilot's Bride, 1880–1883

I HAVE OFTEN BEEN ASKED THE REASON WHY I DID NOT WRITE out my shipwreck experience at Kerguelen Island of eleven months and five days. The answer was that I had lost all of my written records of that voyage, as my journal was lost with the *Pilot's Bride*. And if I should undertake to write the account of that ill-fated voyage, I should have to write from memory and it is hard to recall to mind all the incidents that happen on a three years' voyage, and that I was not qualified or had the talent for such a task. But in the end they have prevailed in my undertaking to write an account of what I can recall to mind of the incidents that happened at that time. And [I] am in hopes that some one will be kind enough to rewrite and revise it for me and for publication if they think it worthy of the trouble.

At the first I will give a short description of the Island of Kerguelen.[1] The most northern part be in Lat. 48°29' S. and is called Bligh's Cap and lies from the mainland [of Kerguelen] about twenty miles E.N.E. The southern point lies at 50°02' S.L. on 68°01' E. and is called Solitary Island, and it is a small low rock. In heavy weather the sea washes over it and it lies from the mainland about twenty-five miles S.S.W. The main island is high and mountainous and the island has many large bays and in most of them there is good harbors where all class of vessels can find good anchorage. As you stand into the land from the northward, the land looks bleak and desolate, with the tops of the mountains covered with snow. But as you draw in with the land, you will see vegetation in green patches and them green patches is wild cabbage and moss and kind of reedlike grass but not a tree or bush of no kind.

On the lee side of the island (that is, from Christmas Harbor to Royal Sound), there is three large bays and in these large bays is

smaller ones. The first of the large ones is White Bay, then
Rhodes Bay, and the next and largest of them all is Hillsborough
Bay. This bay runs inland about forty-five miles, and the bay
about cuts the island in two, and it has many bays branching off in
different directions and is full of islands. It has some of the finest
harbors in the world. There is also many smaller bays between
Christmas [Bay] and the Royal Sound. Some of [these] are Foul
Hawse, Muscle bays, Cumberland, Breakwater, Morgan's,
Eclipse, and Tyzack bays.[2] The land around these bays is mostly
high and bleak looking and is composed of dark rock with now and
then streaks of grey granite and hard blue stone. From Tyzack
Bay about twenty miles S.S.E. is Kent Island and the Rocks of
Despair and inshore of them seven miles S. by W. is Betsy Cove
or the whaleman's Pot Harbor, and it is a fine small harbor and is
land locked. It has some low land around it with small hills and
valleys with high mountains in back. From Betsy Cove, the land
on the seashore is low and rolling.

After leaving Betsy Cove, running to Cape Digby, you will see
the land dotered [that is, dotted] with albatross and one of the
most conspicuous objects is Mount Campbell. It rises like a tower
the height of five hundred feet or more, and when seen at a distance it looks like a ship under topsails, and it can be seen long before the lowland that surrounds it. As you round Digby Cape the
land [is] low, stretching back to the mountains in back of Betsy
Cove to the distance of thirty miles, and the land is wet and
marshy and full of small lakes. There is one large one at the head
of Royal Bay. It is about fifteen miles around it and it is full of
small islands, and the stream is about twenty feet in depth and
about thirty yards in breadth, and the distance from the lake to
the sea about [a] mile and [a] half. In the spring of the year it must
throw off [a] large body of water for there is a large sandbar at its
mouth. Royal Bay [that is, Royal Sound] is an open bay and not
[a] safe place to anchor in. From this bay is lowland wet and
marshy until Shoal Water Bay comes. Along the coast the water is
shoal from a mile to two miles from the land, and this coast has [a]
large body of kelp or seaweed that grows to the length of two
hundred fifty feet and measures around [a] half-inch and full of

long tapering leaves and small air balls of an oblong shape.

On the west side of Shoal Water Bay the land is high at the water and is called Prince of Wales Foreland and it forms the N.E. side of Royal Sound. Kerguelen Head on the S.W. side of this sound is of good length, running into the land for twenty-five miles and it has some fine harbors and is full of islands, these islands forming and making good harbors. Some of [the] principal harbors are Three Island Harbor, Malloy Point Harbor, where the Americans had their transit of Venus party in 1874, and Observatory Bay, where the English had their party at the same time [see part I, n. 19]. On the west and south west side of the sound the land is high, some of the mountains running to [a] height of three and four thousand feet. But Kerguelen Head is the most conspicuous landmark there is, it being [a] high peak and rugged looking and looks as if it had stood many a storm. From Kerguelen Head to Cape Bourbon, a stretch [of] about ninety-five miles, [the] land is high and mountainous. On this side of the island there is the highest mountain on the island as it has an elevation of about seven thousand feet and is always covered [with] snow and is called Mount Ross. At its foot is a glacier that comes out at Big Half-Moon Beach. On this side of the island there is four [*sic*, five] bays, Greenland, Swains, Iceberg, Table, and Sprightly bays.

In and on the outside of these bays are many sand and rocky beaches where there is some background, and on this background there grows a kind [of] barn grass that grows about [a] foot in height. On these beaches the sea elephant come up to have their young and to shed their hair in their proper season, which will be explained in the body of this work. Cape Bourbon is the most southerwesterly point of the main island from that point to Cape Français or Christmas Harbor; [on the] west and southwest side of the island there is but two bays, Melissa's Bay and Thunder Harbor Bay. The land that divides these two bays is west island and between this island and the mainland is a narrow passage called Marienne Strait. Vessels can go through or find good anchorage in these two bays and on the outside of them are [a] good many beaches, fine for sea elephant. [On] some of the beaches it is impossible to make a landing on them on account of the heavy breakers, and the land [is] bleak and desolate looking with but little vegetation to be seen and the tops of the mountains are always covered with ice and snow.

As you look to the head of Thunder Harbor you can see an iceberg [that is, a glacier] that comes down to the water and then rises perpendicular to the height of two or three thousand feet, and when the wind is to the north and east there is one continual fall of ice. This iceberg extends across the island in a southwesterly direction coming out on [the] Iceberg Bay side of the island, and it must have several branches from its centre for the iceberg comes down to the water at the head of several of the bays on the lee side of the island. From Marienne Straits to Christmas Harbor the land is high; no lowland to be seen. Northwest from Christmas Harbor there lies a group of small islands about fifteen miles [off], called Cloudy Islands; more about them further along. The weather at this island is cold but not so freezing cold as we have in the same latitude North. Such is the island where the schooner *Pilot's Bride* was lost and where myself [and] twenty-one men lived eleven months and five days of privations and disagreeableness. And [I] hope it will give the reader some pleasant hours in reading the account of this ill-fated voyage.

In June the year of 1879 I arrived at New London from Kerguelen Island from a voyage of twenty-two months and I received a kind reception from my owner, Mr. C. A. Williams. After the first greeting Mr. Williams said, "Well, captain, I suppose you want to stop home this coming winter."

My answer was yes, sir, that that was the understanding when I sailed last voyage, and he said, "Very good, captain, and during the time you are at home look out for a schooner that will suit you and I will buy her and we will have her all ready by the time you wish to sail in the spring." That was something unexpected by me.

After being at home three months I began to look about [for] a schooner, but could not hear of one that suited me. In November I advertised in the Boston papers for a schooner of about one hundred and seventy-five or two hundred tons. From that time until we bought it, it seemed to me that every one that had a

schooner in the United States had her for sale. But the most of
them was old traps and them that was good was of too much ton-
nage for me, for I had a look at all of them and found none in the
New England states suiting me. In January [I] heard of two in
New York and I went and had [a] look at them, the *William Knight*
and *Pilot's Bride*, and I found that the *Pilot's Bride* filled the bill so
we bought her, and we got her at a good bargain she being a fine
strong schooner of about two hundred tons and needing no repair-
ing. All she needed was some alterations from a trading vessel to a
whaling vessel, such as cutting-in gear and boats and the outfits in
general of a whaleman that is fitting for two years' voyage.

Some time in March I received a letter from Mr. Williams re-
questing me to meet him at New London as he was going to California and did not expect to be again in New London until after I
had sailed. I was living at that time in Danvers, Massachusetts.
The next day after receiving his letter I went to New London and
met Mr. Williams at his office. After greeting me he said, "Captain, I have sent for you so that I can get your plan and ideas of the
coming voyage for I expect to be away from New London for
some time and you will sail before I get [back] from California,
and while we are about it we had better set the day of sailing so
that we can have ample time to get the schooner ready for sea."

"My plan and ideas is, Mr. Williams, if you approve of them,
to fit the schooner for twenty months with about two thousand
barrels of cask³ for oil and salt for three or four thousand sealskins
and let me get away from New London about the twenty-second
of April so that I can take the first of the elephant season at Kerguelen Island — that comes about the first of October — so that I
can take all the oil I can by the first of November or the middle of
that month and from that time I will give up elephanting until the
right whaling season comes in. That will give me good three
months to look for that seal rookery that I think there is somewhere on Kerguelen Island."

"But, captain, what reason have you for thinking there is a seal
rookery of three or four thousand seal and why has it not been
found before this? The island has been worked these many years."

"Mr. Williams, one reason is this: oil has been fetching a high price. One barrel of oil was worth three sealskins and no one liked to give up a certainty for an uncertainty, therefore no body has looked for seal. But things has changed about and you know for the last number of years we have been getting from seventy-five [to] one hundred a year in the shedding season."

"Well, captain, I think your ideas of the seal very good, but I think your estimation large of the number of seal. Why, I shall be quite satisfied if you get on the voyage two thousand barrels of oil and five hundred seal skins."[4]

"Well, Mr. Williams, I shall not be satisfied with five hundred seal if I can find the seal rookery and I have an idea there must be a seal rookery within a hundred miles of Swains Islands where I got them last year, to [that is, since they] throw off so many young seal, and if I can find it I shall find a large number of seal and with your permission I will have a look for it and I think I shall find it."

"Captain, why not fit the schooner only for seal, if you think you can find the seal rookery, for sealskins are fetching a very high price."

"Mr. Williams, I think we had better have two strings to our bow. If one should fail we shall have the other one to fall back on and I can not afford to lose what I have in the *Pilot's Bride*, which I should do if I should fail to find the seal rookery for you know that I have got every dollar that I have in the world in her. It is not because I am afraid the seal is not there, for I am confident there is seal at Kerguelen Island, but I want something that I can fall back on if anything should happen that I am unable to find the seal and I can be taking oil up [to] the last of November for the sealing season does not come on until the middle of December."

"Captain, now about getting home your oil. The schooner will not carry two thousand barrels of oil. What is your plan about getting it home?"

"My idea was if I get one thousand barrels of oil or more after right whale season is over — that will be in May — I will take it on board and run to the Cape of Good Hope and ship it home and if I find the seal rookery I will ship the skins that I get to England at the same time and then return to Kerguelen Island for the next season."

"Captain, your plans and ideas of the voyage is very good and they meet my approval. How long do you stay in New London?"

"I shall return home tomorrow if there is nothing to keep me from doing so."

"Well come to the office before you leave for home."

The next day I went to the office but found Mr. Williams on the wharf and he said, "Captain, I suppose you are about to leave for home?" I told him that I was.

Then he said, "Captain, I am looking at the *Pilot's Bride.* Is she not a fine schooner? Do not let them make no alteration in her mast and [I] think you had [better] be in New London by the last of this month so that you can overlook things and to see that they do not forget to put in [all] that you may be in most need of on the voyage. I will leave orders to put in enough salt for thirteen hundred sealskins. And Captain Fuller," looking at his watch, "it is about time for you to catch your train. There is only one thing more to say: you have [a] fine vessel and if you do not have everything that you need for the voyage it will be your own fault; write us whenever you can and may God speed you on your voyage and at the end of twenty months may you safely return home with a good catch of sealskins and oil," and with a strong grasp of the hand we parted. Little did Mr. Williams or myself think when we were bidding good-bye that it was not for twenty months but that it would be thirty-five months before we should meet again.

The last of March I took my family and came to New London and there I remained until I sailed on April the twenty-second. One day I met Mr. Lawrence, one of the firm of Lawrence & Co. He stopped me on the street and greeted me by saying, "Well, Captain Fuller, Williams has got a fine schooner for you and I suppose you are going to Kerguelen Island."

My answer was, "Yes, that is where she is going."

"Well, captain, we are thinking about fitting [the] bark *Trinity* for Heard Island.[5] What is the prospect do you think of her getting oil at that island?"

"Mr. Lawrence, I do not know much about Heard Island, but I think she ought to do well for the island has not been worked for some time. Who have [you] got to go [as] Master of the *Trinity*?"

"Captain, John Williams is expected to go as master of her."

"About what time, Mr. Lawrence, do you expect the *Trinity* to sail?"

"We do not expect to get her away not until the last of May or the first of June."

"What schooner are you going to send as tender to the bark *Trinity*?"

"She is going on her own hook."

"But Mr. Lawrence, Heard Island is a hard place to work without a vessel to tend on the bark and I should not care about going there alone. You know there is no harbor that a vessel can lie [in] with safety."

"Captain Fuller, the bark *Roman*[6] got her voyage there once without a tender and what has been accomplished once can be accomplished again."

"Yes, Mr. Lawrence, that is all right but you must not forget that all the bark *Roman* had to do was to go to the point and haul off her oil at last end of the season. At that time of the year you have the best weather, but with the *Trinity* it will be different. She will get there in the spring of the year, the worst time of the season to land her cask and provisions on the point and a nasty job it is. I suppose that Captain Williams will only work the Point [and] Whiskey Bay?"

"Yes, them two places; but Fuller, how be you going to get home your oil? The *Bride* will not carry all you get in twenty months for you got eleven hundred in one season last voyage."

"Well, Mr. Lawrence, you say you are going to fill the *Trinity* and if you do I will send home my first season catch by her if she is not full; but if I get one thousand barrels of oil or more I shall run to the Cape with it."

"Well, Fuller, suppose that some thing should happen that the *Pilot's Bride* should get lost? How are you going to get home?"

"Well, if she should get lost at Kerguelen Island and I save my boats, I will get down to where the *Trinity* will make her headquarters, that will be Three Island Harbor, I suppose, then come home in her if Captain Williams will give me and my crew a passage in her."

"Well, Fuller, suppose that something should happen to the *Trinity* while she is at Heard Island? Will you go over and look for her and do what you can and take them off?"

"Mr. Lawrence, as I told you before, If I get one thousand barrels of oil, I shall leave Kerguelen before the *Trinity* gets over, but if anything should happen to the bark *Trinity* most undoubtedly I will go over to Heard Island and get Captain Williams and crew and give them a passage to the Cape. I hope that you do not think that I am as mean as all that, that I would leave him on Heard Island knowing that he was there and come home without my going over and seeing what I could [do] for him, for I should like to have him do as much for me."

"Where do you make your headquarters, Captain Fuller? And if any thing should happen to you, Captain Williams will look out for you if you will do the same by him."

"I will agree to that, Mr. Lawrence, and my headquarters will be in Norton's Harbor." From that time up to the time of my sailing, whenever I met either of the Lawrences they would always speak of my going to Heard Island, but [I] thought nothing of it at the time but had cause in the end to remember them, as the reader will see. About a week before I sailed, I met Captain Williams on the street in New London. In my conversation with him, I told him what the Lawrences wanted me to do and what I had agreed to do. He said they had spoken to him about it. "But, Captain Fuller, if anything should happen to you while you are on the windward side of the island, I can not come to where you are in the bark and if you should lose your boats you will have to stop there until they send some one from home to get you and your crew."

"Well, Williams, it is a one-sided game that you want to play with me. You can come to the windward just as well as I can come to Heard Island. But, Williams, there is one thing about it. I will not come home and leave you on Heard Island if anything should happen to you."

"Captain, [as] you know the bark *Trinity* is not a schooner, and another thing, I am not acquainted with that part of the island."

"Williams, it makes no difference whether you are acquainted or not. You can come and have a look for me and when you get [in sight] of the bay where I am I will get off to you [one] way or another. As for the bark, she is a good worker and sails well. She can work to windward as well as the *Pilot's Bride*. But, Williams, you can do as you have a mind about looking for me if I am not at Norton's Harbor when you are ready for home." That was all the conversation that Captain Williams [and I] had about the two voyages for it was not very likely that I was going to tell him that I though there was seal at Kerguelen Island.

A day or two after my conversation with Captain Williams I met Mr. Lawrence. He said to me, "Captain Fuller, what are you taking so much salt for?"

"Well, Mr. Lawrence, we have been taking some seal from off Swains Islands for years, and I am in hopes to get a hundred or two this voyage."

And he said, "I hope you will, Captain Fuller, but I do not have much faith in Desolation for seal."

The 22 of April [1880], the day of my sailing, came at last. It was a beautiful day with a good breeze blowing from the west north west and clear sky with some flying clouds. At nine o'clock I took leave of my family with the expectation with God's great blessing that at the end of twenty months we should meet again, and I went on board of the schooner *Pilot's Bride* and took command of her. The mate, Mr. Chipman, had been on board some time before me and had hove short on the chains and hoist the sails. The pilot coming on board I gave the order to get underway, and it was but a little while before the *Pilot's Bride*'s head was pointed to the sea. As we went by Fort Trumbull, they dipped their colors and we returned the compliment and sped down the sound toward Montauk Point. At noon the pilot and the two or three visitors that had come on board to see us off and then to return in the pilot boat to New London left us inside of Montauk Point after bidding good-bye.

The officers and crew went to work getting the chains unbent and stowing the chains in the chainlocker and getting all movable things secured before getting outside in rough water. Along in the afternoon the wind became light but not before we had made a good offing from Montauk Point from which point I took my departure steering southeast. At five o'clock, having got all secured, [I] got all hands on deck and chose watches and boat's crews. The officers of the *Pilot's Bride:* E. Chipman, first mate; Mr. M. Fuller,[7] second; T. Gray, third mate; and P. H. Glass, cooper and carpenter; and two boatsteerers, C. Odell and G. Manice; steward, A. Manwaren; and cook, J. Thompson; and seven men in the forecastle; and I was [to] stop at the Azores and Cape Verde Islands and get Portuguese and make the crew twenty-six men all told.[8]

At sunset Montauk Point was low down on the horizon, and as darkness came on thoughts of home and of loved ones would come in my mind: when shall I see you again and them that are more dear than all the world to me? How many of the sixteen will see you, old Montauk, and what trial and suffering have we got to endure in the South Sea before we shall see you again? These thoughts would come in my mind as I watched Montauk light growing dimmer and dimmer in the distance and at last vanishing from sight.

That night and the next day the wind gradually backed around to the east south east and [at] four o'clock in the afternoon it began to blow a strong gale of wind and kicking up a heavy sea. We kept shortening sail until the *Pilot's Bride* had nothing on but a two-reef foresail and a storm trysail, but she behaved well. About seven o'clock she shipped a sea and carried away the faring to the galley, and it went into the lee scuppers, and the cook happened to be in it at the time and when he was coming out of the galley he looked like a drowned rat coming out of his hole. The first thing he said after looking around was, "Aye, mon, I thought I was overboard by faith," and I said to him, "Doctor,[9] you are better than a drowned man; you must not be afraid of a little water."

"Aye, captain," he said, "It's not of a bit of water that I am afraid, but if there had been another come on board before I got out of the galley what would have become of Thompson the cook?" And as he was speaking she shipped another sea and cook seeing it coming tried to get out of the way, but as he turned the sea struck him full and by in the backsides and end over end went poor cook again into the lee scuppers floundering about like a porpoise until Mr. Gray went [to] his assistance. For about a month he was chaffed unmerciful by the crew.

That night about midnight it began to moderate and next morning at sunrise the wind came out west north west and fresh, and at noon we made all sail and kept away for the Azores Islands, the crew employed standing masthead looking for sperm whale, the officers and boatsteerers fitting their boats with whaling gear. We had beautiful weather from that time and until we left [the] Azores. About the first of May we sighted the island of Flores, one of the northernmost islands of the Azores group.

The next morning I took a boat and crew and pulled in shore to a place called Tiesh, but the custom house officer would not let me land not until an oldish man came down and had a talk with the officers. After a little trouble about the schooner's papers, I was permitted to land. After landing I found that oldish man was a priest and as soon as I had landed he took me in his charge, asking me to [his] house; and a fine house it was and well furnished with American furniture and he having a good number of servants about his house. He wanted me to let him do all of my business — the getting of my potatoes and the getting of the men. I told him that he might but he would not do nothing until after dinner.

There was some Portuguese that could speak good English and I asked them about the priest. They said that he was a good man and that he had more to say about the island than the governor, and I think what they said was true. After a little while he gave out to the people that I wanted so many potatoes and eggs and fowl and he setting the price that I should pay for the things, but he stipulated in the bargain that I should [pay] him the money instead of the people that sold me the things, and I had no reason to find fault with his bargains. He gave me an excellent dinner of roast chicken and roast beef and some of the most excellent wine that I ever drank. After dinner he took me for a walk and showing me all the sights there was on that part of the island and gave me the history of every thing that was of interest, he being well informed in the history of the island, and I was so taken up with his conversation that the time slipped by before I knew it, so we retraced [our] way back and when we got back to the house the sun was low in the horizon but the priest had left orders to have the things at the landing at such a time with eight young men and they had carried out his orders.

The priest accompanied [me] to the landing and on the way he stated to me that one of their shore boats was going off to the schooner and that his assistant was going off in her, and if I had a little tobacco to spare he would be very thankful for a little. I promised him that he should have some or anything that he wanted, but he would take nothing but some good American tobacco and he did not want the custom house officer to know nothing about him having the tobacco. The priest gave me his blessing and offering up a prayer to God that he might bless me in all my undertakings. After the prayer I bade him good-bye and I got on board long before the shore boat, and when she got alongside I had the tobacco put up in a parcel for the priest and some for the others and the custom house officer did not refuse good American tobacco.

At dark kept away upon my course to the southward; the next day in the afternoon sighted Fayal and Pico islands, but did not stop but kept on for the Cape Verde Islands. About a week after [our] leaving the Azores, the wind came southwest dead-a-head. For ten days we could do nothing but lay first on one tack then on the other and let the wind blow itself out. After that we had fine weather and we arrived at Brava, one of the southernmost islands of the Cape Verde [Islands], where I got the remainder of my crew — four men, making my crew twenty-eight all told.

Three days after leaving Brava, we fell in a school of sperm whale. We got one, the wind being very light so that we could not follow the school. But we was thankful for one and it made us sixty-five barrels of oil. We had plenty of rain while boiling the whale out for we had got out of the northeast trades into the variable winds that lay between the northeast and southeast trade winds.

Some little time after getting the sperm whale, one of the men was taken sick. Timothy Reardon was his name. He broke out in sores from the top of his head to the bottom of his feet, and I thought that we was going to lose him. He was a fearful sight to behold. He grew so poor that one could count all the bones in his body and [he was] so weak that he had to be lifted from his bed and one man [had to] tend on him all the time. I think it must to have been blood poison and he must have [gotten] some of the blackskin from the whale blubber into a cut he had on his hand.

One fine day I had him carried on deck and we fitted up a bed on the forehatch and laid him down on it and made him as comfortable as we could. In about two hours one of the men came and said that he thought that Reardon was dying, so I went along to where he laid but he seemed to be about the same. While I was looking at him all at once his breathing seemed to stop and the mate came along and looked at him and he said [to] me, "Captain, damn it, he is dead is he not, and I think we shall have to chuck him overboard to the fish."

He got through speaking and I was turning to go aft for something when Reardon opened his eyes and in a weak voice said, "No, you will not have to throw me overboard, Mr. Chipman," and from that day he began to mend and in about a month's time he was as well as any man on board.

We was about a week working across the line into the southeast trade winds. When we did get them we had them strong, the *Pilot's Bride* running off her two hundred and twenty and thirty miles a day until we got within one hundred miles of Trinidad [Trindade] Island. Then we had it calm for fourteen days, not a breath of wind and the sea as smooth as glass. Now and then a cat's-paw for a minute then calm again. But we were not alone, for we could look in all directions and see vessels in the same predicament as ourselves and we took some comfort in the old saying that misery likes company.

One morning I found that we had drifted through the night within a mile of a large ship and after breakfast it still being calm I told the mate that he might go on board of her and give the captain my compliments and ask him for his position. He took his boat and crew and went on board of her and spent the best part of the day. On his return he gave me the name of the ship and her position; we was about the same in time. Mr. Chipman said the captain wanted to know where [we] was going and what we was going to do and where abouts Kèrguelen laid. I expect that Mr. Chipman gave him the history of the voyage as he had the gift of gab.

On the fourteenth day of calm, the weather took a change. The change came in a squall of rain. I thought that I had seen it rain before, but all the rain squalls that I had seen was nothing but scotch mists to this, for it did not come in drops but seemed to come in streams and it lasted about [a] half-hour. When it cleared off we did not lack for wind, for we had all we wanted for about twenty-five days, which carried us down to Crozet Islands when we had light head winds and calm for three days. Then [the] wind came out strong from westward the next day. [We] had [a] strong gale. We put the schooner under short sail at night, and, at ten o'clock when I went below for the night, I gave orders to Mr. Gray to give me a call if the weather got worse, for the schooner was at the time making good weather of it and shipping but little water.

About eleven o'clock as I was laying down on my bed, I heard [a] crash. I sprang from my bed and jumped on deck. As my head came out of the companionway, I just had time to see that the port waist boat had gone and as it came aft and got under the port quarter boat it rose on a sea, struck the quarter boat, and carried away boat, davits, and all the tackling that belonged to the boat, making a clean sweep of the port side. As soon as I could take in the situation, I had all hands called to shorten sail.

It was blowing heavy gale and a tremendous sea running. When the schooner would get on top of a sea, she would shoot from the top like shot from a gun. As we got the mainsail in and [was] furling it, there came along another heavy sea and cut the stern of the boat that we had across the stern; that left us with only one boat. We got the storm trysail and put the reefs in the foresail and hove to heading to the northward. For about twenty-five hours it blowed a hurricane and the sea looked like mountains coming down upon us but the *Pilot's Bride* made good weather of it by shipping but little water.

At the end of the second day it began to moderate and the next morning [I] kept off on my course and we had no more storms on the rest of the passage that we had to lay to for. The day before we got to Kerguelen Island I had it calm and the sky was as clear as a bell and the sea was smooth as a house floor. At noon I got my latitude and at two o'clock I got the longitude and it put me just two hundred and eighty miles from Cape Français. A little after noon it began to breeze up and before dark it was blowing strong, but the water kept smooth all through the night. The next morning the sky became overcast with flying clouds and strong winds. We had to take in gaff topsails but we could carry all five lower sails. About noon it became cloudy, and once in a while the sun would break through the clouds, and as I was getting the sun I thought that I could see the land as the clouds would lift and the land looked to me like the Cloudy Islands. I could only get a glimpse of it before it would shut in again, so I hauled up another point so as to go within five miles of it, but we ran by it and did not see it. We must have sailed between the Cloudy Islands and Bligh's Cap, for we was keeping a sharp lookout for land.

About three o'clock in the afternoon the cooper happened to look astern on the port quarter and saw Bligh's Cap off about fifteen miles, so we had to haul on the wind so as to fetch into Howes Foreland, that being [the] only place that we could get into as it was blowing almost a gale of wind. It was of no use of trying to get in some of the other harbors, as they were all to windward of me. So I told the mate to get one anchor ready and that we would get the other out when we came to anchor. When I hauled the schooner on the wind it seemed to me as if it was blowing two gales of wind in one, but the schooner was stiff as a tree when the heavy puffs would come down from off Christmas Harbor Head. She would hardly heel to them but she did make the water fly from her stem to her stern and half-way to the masthead.

At six o'clock I came to anchor in Rhodes Bay, about three miles from Fuller's Harbor. We laid there three days, it being thick fog in the meantime. While laying there one afternoon I took the boat out and went on shore gunning. I got some teal duck — they are about the size of a pigeon and of a dun color, their flesh is a fine flavor — and about a dozen of young albatross. The young albatross is fine eating, the flesh looking like veal but the taste much nicer and its flesh more firm. I also got half of a boat load of wild cabbage, which gave us quite a fresh supply of meat and vegetables. From Rhodes Bay we went to Norton's Harbor, my headquarters, and went to work breaking out the schooner and landing all of the spare provisions and shooks.[10] I kept in the schooner about twelve hundred barrels of cask and five months' provisions to go to the windward with. While we was doing that the carpenter went to work on the boat. We got stove coming down and when it was finished the boat was as good as new. By the time the boat was finished the schooner was all ready to go to the windward of the island.

About the twenty-fifth of August I left Norton's Harbor for Young William's Bay on the west side of the island. As we got outside of the harbor one of the men up aloft doing something reported a sail in the direction of the Rocks [of] Despair. As it was blowing at the time, I did not run off to see but went into Morgan's Bay and came to anchor and let the crew go on shore after rabbits and when they came on board at night they brought about two hundred rabbits with them. We only stopped overnight in Morgan's Bay as the wind came around to the southwest in the night. At daylight I got underway and as we worked along the bays we kept picking up sea elephant and sea leopard.

This sea animal is about seven feet in length and from three to four around, and of a dark bluish color on the back and of a gray white on the belly and they are dotted with dirty yellow spots all over the body, and their head is longer and slimmer than the elephant is and the nose looks more like a sheep's nose and their tongue is rough like a bullock's tongue. They do not come up on the sand beach in large numbers the same as the elephant but haul up two and three on a rock that the tide washes. They generally come up when the tide is falling and stop up until the tide rises again and washes them off the rock. We take them for their oil, it making very clear white oil, but I have taken some of the skins home with me and had them tanned. They make a fine grain leather for the tops of shoes and I think that the leopard skin could be made of some value if they were properly put on the market.

There is one thing curious about the sea leopard. No one

knows nothing about them, where they come from or where they have their young as I have never seen their young with the old ones. They come around Kerguelen Islands in May and stop until October and they then disappear; you may see one now and then. They are carnivorous animals as they will catch penguins and lie in wait for sea birds and also catch fish. They are very quick in the water and when they are attacked in the water they will turn and attack a boat and seem to have no fear, but when on the rocks it is otherwise. I have had them turn upon me when I have struck them in the water; they seem to know where to take hold of the boat by putting their jaws on each side of the keel and crush the planks in like an egg shell with their teeth, and their teeth is something like the dog's in front, but much larger. The back double teeth has three prongs on top and are solid.

That night we came to anchor in Christmas Harbor. This harbor is the last on the northeast side of Kerguelen; it lays just inside of Cape Français. When you are coming from the southwest, Cape Français looks like a large barn with a large cupola on the top of it. On the west side the land is high, with an overhanging precipice that looks as if it might fall into the harbor. At the mouth of the harbor is a point running across about half-way with a peculiar looking arch on it and is called Arch Point. When you are at anchor inside [you] can look through it and in the season it is full of penguins. There is very little vegetation, the land being bleak and wet and on both sides of the harbor the land is high, running up in rock ridges to the height of two thousand feet, and up on the west side where the overhanging precipice is and close to it you can get petrified wood.

At Christmas [Harbor] it has been the custom to stop until we get the first of a norther and then work out to the Cloudy Islands and from them run to Cape Louis or Marienne Straits, and we usually get there before bad weather comes on. The worst of going to the straits is thick weather as the two harbors that lay between Cape Français and Cape Louis are not good bays to get into in thick weather. The name of them is Rocky Bay and Soulskin Bay. The land being so high and on the northwest side of the island that the least fog it banks up and shuts in the land, and we dare not run close in on account of the heavy sea that is always setting in shore, the land forming a big bight, and it is getting across this bight where the trouble come in.

I laid in Christmas Harbor about a week. Most of the time it was blowing a gale of wind from [the] northwest. At last the wind became light from [the] southwest, so early in the morning I got underway and the southwest wind just took me within a mile of Cloudy Islands. Then the wind worked around to the northeast, and by noon it became strong, but no fog with it, until I got within two or three miles of Cape Louis, and then not so thick but I could keep run of the land, and it kept growing thicker as I came up with the cape. As the tide was setting to westward I went through on the northeast side and came to anchor in Marienne Straits, it having good anchorage as you come from the northeast. Soon after coming to anchor it became thick and raining and blowing heavy for two or three days.

The straits is a narrow strip of water about one mile and [a] half long and [a] half-mile in width and it cuts West Island off from the mainland. On the northeast side of West Island is Thunder Harbor Bay and on the southwest side is Melissa's Bay. At the head of this large bay is Bull Beach and about four hundred yards before coming to Bull Beach on the east is Young William's Bay. You can see all of the bays and beaches from the straits. As you look around from the straits of a fine day you will see nothing but bleakness and now and then small patches of cabbage and moss in sheltered spots, the land being very cold and wet. Looking to the southwest toward Cape Bourbon and it happens to be clear and [if you] should look over Hell Gate, you will see the volcano mountain two or three thousand feet high with its black smoke rising high in the air, and all up the mountain side ice and snow — nothing green, all bleak and desolate as far as the eye can see — with glaciers coming down to the beaches at the head of all the small bays and some of the glaciers coming to the water's edge. In going out the southwest passage it is very dangerous, as you go between two reefs, and if the wind should drop off, a vessel cannot turn about in it and there is always a heavy sea running into the straits, and on the northwest side of the passage the land is high and in a norther you are apt to have it calm in the passage.

I got about fifty barrels of oil in the straits and soon as the wind came to northward, which happened to be in the night, and soon as the tide was running through to the southwest I got underway. It was blowing a fresh breeze right out of the straits. After getting underway, I had the gaff topsails loose all ready to sheet home if [I] should need them. As I drawed down to the mouth of the passage, the wind freshened up strong and shot me into the passage about half-way and then left me becalmed, the sails hanging loose along the mast and booms. I had the gaff top-sails sheeted home and they being so large and high that they caught the wind when the lower sails would be becalmed. It was slow sailing against the heavy sea that was rolling into the pas-sage, but at times a little puff would fill the mainsail for a minute and help out the topsails. It was [an] anxious time to me if it was only for [a] short time but it seemed to me as if I was two or three hours getting out of the passage when in fact we were only about twenty minutes going through when the breeze struck us in Melissa's Bay.

I had the gaff topsails taken in as the wind was freshing. There was heavy sea running in the bay, showing all the breakers and reefs — and a dangerous looking bay it was. As I sailed down the bay I could see that the wind was blowing out of Young William's Bay. I had every thing ready so that we could take in sail at a min-ute's warning and the anchors ready to let go. I hugged the north shore as near as I dared and when I got to the point of Young William's Bay I shot as far into the mouth as she would fetch. I had the anchor let go and the sails taken in; then we warped the schooner in with a kedge to the anchorage.

Young William's Bay is a bar harbor, the bar extending into the harbor about four hundred yards, and has about three fathoms of water on it at low tide. We anchor over the bar in twelve fathoms of water, with the large anchor up the bay, and give a scope of sixty fathoms of chain to ride by, as we have the strongest wind down the bay. Then we put the small anchor off shore with

forty-five fathoms of chain to keep her from fouling her anchors.
[At] Young William's Bay, the land is high on both sides but that
on the north is highest and the formation is of a volcanic nature
rising in terraces of a rotten blue stone and rugged precipices the
height of two thousand feet and grey with age with no vegetation
but little green patches of moss on the south side. Towards the
head of the bay the land is of the same formation, except it has a
rotten red stone mixed in with the blue, and toward the mouth of
the bay the land gradually slopes off to about five hundred feet
and is wet and marshy. At the head of the bay the land is low, extending inland about two miles and gradually growing narrower
as it meets the rough mountain in back. The land on both sides of
the bay being high and that at the head low, it forms a funnel and
when the wind is to the north or northeast the wind comes down
the bay with terrific force, picking the water up and making the
bay one sheet of white water, and a man can hardly stand up on
deck in some of the squalls.

About four hundred yards from the mouth of the bay is Bull
Beach, laying between two mountains. One of the mountains
separates [it] from Young William's Bay and the other from Melissa's Beach. It is a sand beach a mile long, except the lowest end
that is rocky and in each end is some grass of the reed kind. The
beach extends inland to [an] iceberg about five miles and it's a
level place dotted with green moss and strewed with black and
yellow pumice stone and cut up with small water streams. All
come from the iceberg except one large stream, which comes from
an inland lake, and all the small streams run into it, forming one
large stream at the lower end of the beach. On this beach I have
picked up the white topaz and that is the only gem that was of any
value that I ever picked up on the island.

After getting the schooner moored the mate and myself went
on shore and walked over the hill to Bull Beach to see what prospect there was for elephant. From the top of the hill with the help
of a glass we could see all the elephant on the beach, and as we
looked down on the beach it was a sight for the eyes, for there was
something like three hundred bull and cow elephant up on the
beach and the bullmasters had formed the cows into pods. Some

of the pods had ten, thirty, and sixty cows in each pod and a
hundred or more bulls lying round on the outside of the different
pods.

As we was looking, one of the outside bulls got up courage and
attacked one of the pod masters. There was a terrific battle between them, they standing up on their hind flippers and rearing
their bodies in the air to the height of ten or twelve feet, looking
like two giant gladiators of the ancient Roman prize ring, fighting
with the rest of the elephant looking on as spectators. They fight
by striking their bodies together and biting [at] the same time with
their formidable teeth, and whenever they strike and bite they
will tear the skin and flesh from their bodies and [you] can hear
the sound of striking each other and their roaring still farther,
making the mountains ring again with their roaring.

After the two bulls had been fighting about a half-hour, the
one that attacked the pod master he had to back out minus a part
of his trunk. But if he had whipped the pod master it would have
been woe to him, for he would [then] have to do battle with all the
rest of the bulls to let them know that [he] was master of that pod
of cows. There is one thing curious about elephant: when two
bulls are fighting the rest of the bulls will not go near the pod, but
calmly look; nor will the cows try to get away, but huddle closer
together, seeming to be waiting to see which should whip. I have
seen as many as twenty different bulls masters of one pod of cows
in one season, but generally one old bull will make up a pod of
cows and hold them for the season against all comers.

The habits of the sea elephant is but little known and that little
is only of their habits while being on shore. The bull elephant is
an animal [that] when full-grown is about eighteen or twenty-two
feet in length and will measure around across the fore shoulders
when fat ten feet or more. The head is about eighteen inches
across the eyes and about two feet in length. When young they
have the look of a bulldog, but when full-grown they have a proboscis about twenty inches in length hanging down over the
mouth, something like the land elephant in fact. When the sea
elephant has shed its hair and before it has grown out again, the
hide resembles the land elephant's very much.

They have two fore flippers two feet in length, with five distinct finger nails. I have seen them use the fore flippers to scratch the back of their head and around eyes and neck, they using the flipper the same as a human being uses his hand to scratch. They have two hind flippers like a seal and when they come on shore they do not use their hind flippers but keep them closed up like a fan. To propel themselves along they use the fore flipper only by reaching forward with the flippers and dragging themselves along.

The color of the hair when they first come out of the sea is of a light dun, but keeps changing the longer they stop on the land. The skin on a full-grown bull elephant from the head down to their foreshoulders is a good inch in thickness and very tough. It acts as a shield to them when fighting.

The cows and young elephant have one thickness all over their bodies. The cow elephant is much smaller than the bull. They will measure from tip of nose to the end of tail flippers about seven feet and when fat about six feet in circumference. They are of the same color as the bull and they do not have the proboscis of the bull. The head is much smaller, having the looks of a bulldog and when not molested they have a mild eye [unlike] the bull whose eyes is wild and wicked and when molested they turn red. The cow is much the quicker in their motion and are very affectionate to their own young and will fight to their last breath to protect them, but will have nothing to do with the other young pups. They know their own and when the young pups get away from them they will call them back by making a croaking noise and the pups seem to know their mother's voice and will return to the side of their mother and she will smell them all over. The young pups when first born and for the first two months the hair is black and as they grow older their hair turns to a white gray on the belly and a blue black on the back and [they] are very playful when young.

The sea elephant do not have ears like the seal or dog but theirs is the same as the whale. It is only a small hole on the side of the head and a knitting needle will stop it up, but their hearing is very sharp. I think when the elephant is in his native elements he can stop underwater as long as they have a mind to do so and that they do not have to come to the top of the water to breathe the air as the whale or seal, for I have seen a bull lay with his head and half of his body underwater for thirty-six hours and never move or alter his position in all that time.[11]

The elephant season commences the middle of September and lasts through October. That is the pupping season but the bulls come up the last of August, going from beach to beach picking out the best place to locate their pod, and when they first come up they are very fat. They can hardly crawl along the beach and are sluggish and docile, but as soon as the cows begin to come up their manner begins to change. They will not allow another bull to come near them. About the middle of September the cows begin to come up to have their young, and as soon as a cow comes on the beach sand the bull will receive her at the water's edge and drive her into his pod. Once they get in the pod they will stop in but the pod master will keep generally about half way between the pod and the water so as to keep an eye on the pod and the water at the same time, and if he should see one coming out of the water he will go for her and perhaps he will have two or three fights over her and if he should whip, into his pod she must go.

About the first of October the bulls will not leave their pods, for they have all the work they can do to keep the rest of the bulls from the cows for about that time the cows begin to have their young. But now and then a bull more adventurous than the rest will steal into a large pod of cows, but he will keep his eye on the pod master and be in clover. But when the old master happens to see him and sounds his trumpet on his trunk, and he will not sound but once, then Mr. Bull will leave on the double quick; perhaps he will get a taste of the master's striking power before he gets away.

About the last of October the cows begin to leave the beach, the pups being large enough to take care of themselves the cow having only one pup at a time. Before the cows leave the beach they take the bull they cohabit by laying on their sides. It is at this time that the outlying bulls close in on the pod, and when a cow leaves for the water all the bulls that she will come across they will cohabit with her. I think that the cows must carry their young eleven months. At the end of the pupping season the bulls and cows are very poor, [having eaten] nothing but live on their fat all the time they are on the shore. By the first of November all the cows and most of the bulls [have] left the beach.

The first of November the young elephant, that is elephant that is from two [to] three years old, come up to shed their hair. They come up generally on a beach that has grass and water holes. They like to lay in the grass where the sun can get at them and to be sheltered from the cold winds, and they like to get into water holes and wallow and roll about in the mud to loosen the hair, and [they] keep hauling[12] until the middle of December. Then the pupping cows come again. They shed their hair and they like the beach that has grass no matter whether it is sand or rocky. Soon after the cows, the old bulls come up. The bulls and cows keep coming up to shed from the last of December until March in the following year.

When they come up first out [of the] sea they are very fat and round, and all the bulls and cows and the young elephant will huddle together and seem to live in peace. When they are shedding they are very docile and harmless and will take but little notice of a man, and by the time they get through shedding they are nothing but skin and bones. They eat nothing from the time they come up until they go off [to] the [sea] again. The fat of the elephant resembles fat pork but much thicker. The fat or blubber of yearling and cow elephant when they first come up out of the sea is about three inches and bulls six to eight inches in thickness and yellow. When boiling out the blubber for oil the blubber smells and tastes like pork and the scrap is soft, resembling pork scraps. We use a press to get the oil out of the scrap. The bull, when he first comes up and in good order, will make from two to three hundred gallons of oil each and yearling and old cows about thirty-two gallons of oil, but of course the longer they stop on the beach the least oil they will make.

The last of September we commenced killing the outside bulls that laid around the pods of cows on Bull Beach; but in the meantime, while waiting for the last of September to come around, whenever we had an opportunity we went to swampy holes across the bay after blubber and on all of the little beaches whenever the

weather would permit. When we commenced to kill on Bull
Beach we never stop for good or bad weather, but keep right along
killing and carting the blubber to the upper end of the beach so
when good weather came and it was smooth we could haul the
blubber off and get it alongside of the vessel. On the upper end of
Bull Beach there [are] large water holes, and into these holes we
put the blubber to keep it cold so that the oil will not run out as
fast as the men can cart it up from the beach. The officers and
boatsteerers generally do the killing and skinning and the foremast
hands doing [the] carting and backing [that is, carrying] the blubber and the men will generally cart it away as fast as the officers
can skin the elephant. We shoot the bulls and then lance them to
let the blood run off the blubber as they have abundant — some
say that they have as much blood as oil.

  After killing the elephant the next operation is skinning, for
they have a hide the same as a bullock. You cut around the neck
and then right straight down the back to the hind flippers; then
you skin down each side of the elephant to the ground. After taking the skin off the blubber is still on the elephant. Now you cut
right down through the blubber until you come to the flesh. Then
[you] cut off [a] square piece [of blubber] to the ground, and after
carting the piece off you must make a hole in the center by thrust-
ing your knife through it, and then [you] continue until all is taken
off. Then you roll him over so that his belly will be up and then
[you] go through the same operation on the belly as you did the
back and not forgetting to make holes in the center of each piece
that you take off. The hole that you make in the center of each
piece is so that you can get a backing pole through and to string it
on the raftline to haul it off the beach and to soak the blubber
alongside of the vessel so as to get all of the blood out of the blubber. If you do not soak the blubber well it will make black oil.

  The backing pole is a hardwood stick pointed at one end and
about two inches through and seven feet long. Two good men take
one pole and put about [a] half barrel of blubber on it. Then each
man will take hold of his end of the pole and pick it up and let the
end rest on each of their shoulders and walk away with all ease
and carry the load either to the water holes or to the seashore or

wherever you want the blubber put. The first going of the pole will hurt the shoulders some, but in a day or two the men will not mind it much, if any. But on Bull Beach we do not use the pole much — only to carry the blubber from the water holes to [the] water's edge when rafting it off.

We use two carts, the beach being so level and free from stone. One of the carts would carry about ten barrels and the other, six barrels of blubber. The carts had broad wheels so that the wheels would not sink into the sand. The bodies was made of light material, with eight men on the large cart and six men on the small cart. They usually got to the water holes a large quantity of blubber in a day. As soon as the weather will permit and [it is] smooth on the beach, we generally stop killing and skinning and haul off the blubber that is in the water holes so as to make room for more.

The way [we] haul off is this: when a good spell of good weather came and [it was] smooth on the beach, we stop all work and all of the crew go on shore and get the blubber out of the water holes and back it down to the water's edge. Two or three men will put it on a raftline as fast as the rest of the men can [carry] the blubber down from the water holes. The raftline is three fathoms long with a eye in one end. You put two pieces of bull blubber [on the line] and then reeve the end through the eye and haul it up tight to form a butt so that the rest of the blubber will not slip off that end. Then you keep on putting on blubber until the raftline is full, and then you make it fast to the butt of another raftline and keep filling raftlines and making them fast one into the other until you get fourteen or sixteen full and in a straight line along the beach.

Then one of the officers, generally the mate, takes a boat crew and goes and gets the boat with a tub of towline and kedge anchor and comes to the beach with it. Then he heaves anchor and comes to the beach with it; then he heaves the end of the towline on shore and the end is made fast to the raft of blubber; then the rest of the line that is in the boat is run off shore, the other end being made fast to the anchor. As soon as the line is all out of the tub the anchor is let go and the boat is hauled back toward the beach about twenty-five fathoms. This [is] the riding scope so that the anchor will not come home and is made fast in the bows of the boat. After getting all ready the men in the boat take hold of the line that is made fast to the raft of blubber and keep a good strain and gather in all the slack line. As the men on the beach haul the raft into the water, if the men in the boat [do] not get in the slack the sea will wash it up on the beach again. After getting the raft off the beach, one boat can tow it [to] the vessel if it is calm; if not, two boats will. While the boats is towing to the vessel the rest of the crew will be getting another raft ready to haul off. So we keep hauling off all day or as long as it keeps smooth or [we] have blubber to haul off the beach.

Generally after getting a hundred or two [hundred] barrels of blubber off the beach, the mate and half of the crew stop on board the vessel to boil the blubber out. The tryworks in [the] vessel is placed a little abaft the foremast and is made of brick. In building the tryworks they put two layers of brick on deck first, and where the fires come there is a open space under the fires which is bricked over which is filled with water when boiling out so as not [to] catch the vessel afire and the vessel has two 200-gallon pots set in this brickwork for boiling out the blubber in. In preparing the blubber for boiling we take off all the flesh and then mince [the blubber] with a knife. The mincing is this: we take [a] piece of blubber on a board and cut it about through on one side and then turn the piece over and cut that side the same, the cuts being about [a] half-inch thick. Then men will mince and boil out from sunrise to sunset about sixty barrels of oil. We use the scraps from the blubber for fuel, it making a very hot fire.

From the time we commenced killing we had very good weather and all things working in our favor. The last of October we was all through killing on Bull Beach and all boiled out and ready to leave Young William's Bay as soon as the weather would permit, we having taken about eight hundred barrels of oil up to that time. The first of November I left Young William's [Bay], thinking of going around to Table Bay in the southwest side of the island; but after getting outside of Melissa's Bay the wind came out from the southwest and I thought the best thing that I could do was to go into Thunder Harbor as I had a fair wind into the harbor. So at ten o'clock I kept the schooner away and at noon I rounded Cape Louis. The wind came out east northeast right out of the Thunder Harbor, bringing the ice with it from the iceberg at the head of the bay, and the ice troubled us about beating into the bay but we had a fair tide that helped the schooner about working. At sunset we came to anchor with two anchors, one to northward and one southwest with forty-five fathoms on each anchor.

The next day the bay was full of ice; some of the pieces was as big as the schooner. The bay was packed so full that we could not get on shore for two days, but we could see a large number of elephant on shore from the vessel. There is two beaches in Thunder Harbor; one [on] the southwest side is call Blueskin and the one on the north side where the vessel anchors is Shoefoot and they are easy beaches to work on account of they being on the inside of the bay. The bay is about a mile and half wide, open to the northwest.

The third day after coming to anchor in Thunder Harbor the crew went killing elephant on Shoefoot, the elephant being shedding elephant, the pupping season being over. We killed and skinned all day and along toward night we rafted what blubber we had killed and got it alongside of the vessel. Next day [I] left one half of the crew on board to boil out [the blubber] and the rest of the crew went ashore killing and skinning. By the twentieth of November I had all of the cask that I had in the schooner full and ready to go to Norton's Harbor. After leaving Thunder Harbor we kept picking up elephant until we came to anchor in Norton's Harbor, where we boiled it out and put it on shore. After boiling out all of the blubber I found that I had taken about thirteen hundred barrels of oil. The first of December I took on board fifteen bags of salt and left Norton's Harbor to look for seal.

The first island that [I] looked at was Swain's Island. I gave them a good looking over. We got about seventy young seal, that is seal about [one] year or two years old but found no seal rookery and in the meantime, [the] wind coming in northeast, I came to anchor in Breakwater Bay, we laying at harbor some time on account of the weather. While laying in Breakwater Bay I got three more seal. The first good day which happened to be Christmas day, it being fine, I got underway and worked out of Breakwater Bay; and on the outside of Breakwater and off the mouth of Cumberland Bay is three little islands. I sent the mate and third mate in to see if there was [seal] on them and if there was to get what they could and I sent the second mate in on the north point of Cumberland Bay. When the mate and third mate came on board they had twelve seal and some of them was old whigs and the second mate got nine young seal. As the wind keep freshing from the northward I came to anchor in Christmas Harbor, and I had to lay there nineteen days before I had an opportunity to go out, it blowing all the time.

It was while we was working up to the anchorage that I had some words with Mr. Chipman about us looking for seal. He said that he did not see the use of wasting our time looking for seal when we ought to be getting oil. I said, "Mr. Chipman you knew when I came from home it was an understood thing that we should look for seal after the first of December and up to February."

He said that we was wasting our time and that he did not believe that there was a seal rookery on the island. I said, "Well, Mr. Chipman, there may not be a seal rookery, but you can make up your mind to one thing — that I am going to have a look at all of the out-laying islands if it takes me until the last of February for it was for seal that I came for and sealskins I am going to have if there [are] any on Kerguelen."

He said, "Captain, I think you are doing wrong. We ought to be getting ready for right whaling and not be wasting our time."

"Well, Mr. Chipman, wrong or not wrong we are going to have a look, so you need not say nothing more about it, and what you have said will not change my mind one way or the other."

The thirteenth of January the wind became light from the southwest and the indications was that we should have a spell of good weather in the afternoon. I told Mr. Chipman to have things ready for an early start in the morning. That night I had a dream. I dreamt that I went to the Cloudy Islands and found a seal rookery and that I got seventeen hundred sealskins. Then the dream turned and [I dreamed] that Captain Glass [13] came to Kerguelen and got thirteen hundred more from the same rookery and then I waked up. I got up and went on deck. The weather was growing better and I went back to bed again and I went to sleep and I had the same dream over again. I got up and looked at the time and it was just two o'clock in the morning.

I called all hands and got underway and the wind being light I was some time working out of the bay. At daylight we was about five miles north of Cape Français, heading for the big Cloudys, but the wind was growing lighter as the day advanced and at seven o'clock I was within a mile of the big lower Cloudy and the vessel had only steering way on. After breakfast I told Mr. Fuller and Mr. Gray to get their two boats ready and put into the island and when they got in to pull up on the back side as near the land as they could go and see if there was seal on that side of the island. If they did not find seal they were to keep on going and look on the other four islands. And I told Mr. Chipman to get his boat ready at the same time and to go in with the other two boats and when he got in shore not to go up on the back but to go up on the inside. I would work close inshore with the schooner and keep run of his boat for I should work up to the passage way between the islands, and if they was not back by that time I would lay aback for them.

The boats got away from the vessel about eight o'clock. They had not left the vessel more than ten minutes before the wind came out north northwest. And [with a] good working breeze I tacked and stood into the island and got [there] about the same time the boats did and [lay] aback until Mr. Chipman parted from the other two boats and pulled in and got three seal. Then he went on a little farther and came to the inner point of the island and landed, and I see him knock down fifteen more seal and I keeping close in shore until he had them skinned. After he got into his boat and was pulling along the shore, I stood along, the wind being [such] that I could make a long leg and a short one along the land. By the time that I got abreast of the passage it was five o'clock in the afternoon and Mr. Chipman had gone across the passage to the two small islands and was out of sight from the vessel. While laying aback there was about fifty seal came up alongside of the

schooner and played about, and I said to Glass, the cooper, that
the boats must have found the seal rookery, he being an old sealer,
and he said he thought so too by seeing so many in the water.

I laid aback until six o'clock, when I began to be anxious about
the boats. I did not know whether it was best to go through the
passage to the other side or not, as I had never been through it before and did not know if there was any obstructions in it or not.
But at last I could not stand it no longer. I made up my mind to go
through. As I had a leading wind I sent a good man to the masthead to keep a good lookout for breakers and rocks and I pointed
her through, and as I opened out on the other end of the passage I
could see one boat. It was made fast [to] the shore but [I] did see
no men. So I keep beating back and forth for two hours. And at
last I see one boat coming out of a small bight between two small
islands. I tacked and stood in towards the boat, and, when the
boat got within three hundred yards of the vessel, I see that it was
Mr. Fuller, and his boat was loaded down and George Manice, his
boatsteerer, could not hold in no longer but had to stop pulling to
shout out, "Captain Fuller, we have struck the seal rookery three
thousand strong by (d——n)."

I hove aback, heading off shore and the boat came alongside. I
asked Mr. Fuller, what luck? He said that he had found the seal
rookery and he thought there was about two thousand seal on the
island and that he had killed about three hundred on the outside of
the rookery and he left Mr. Chipman and [Mr.] Gray to finish
skinning them. I told him to throw the one hundred and fifty that
he had in his boat on deck and take his boat out of the water. After
Mr. Fuller [had] taken his boat out of the water, I told him to get
ready with his boat crew to go on shore that night and I gave orders to the cooper to get up a barrel of pork and two hundred
pounds [of] bread and the same of flour and I told the steward to
put up some small stores and have them all ready by the time that
Mr. Gray came alongside.

By the time that Mr. Fuller was ready, it was dark, but it happened to be a moonlight night and clear. His boat was loaded
down with his [supplies] and the men, clothing, seal clubs, and
tent poles and beaming boards,[14] and twelve bags of salt. I told

him to kill so as to make up with what I had about twenty-two hundred if he could get them and if the next day was good I would send in the boats in the morning but would go to Norton's Harbor after the rest of the salt when it came in bad weather and that I would be back again as soon as the weather would let me get back.

In about ten minutes after the boat had left the schooner, Mr. Gray and Mr. Chipman came on board. Mr. Gray throwed what skins he had in his boat on deck and I sent him in with the remainder of the provisions and the other articles that Mr. Fuller could not take in his boat. It was exactly 10 o'clock when the boat got back, and I hove to for the night, making all day of it. We got dinner and supper in one. While getting supper, Mr. Chipman asked me how many skins I was going to take. I told him. I gave Mr. Fuller directions to kill enough with what we had to make up twenty-two hundred, but Mr. Chipman thought it too many skins for the salt that [we] had. "You know," he said, "that Mr. Williams said that he would put salt in for thirteen hundred and I know you do not want them to spoil on our hands before we get to the Cape."

I said, "It looks to me, Mr. Chipman, as if we ought to have salt enough for twenty-two hundred, but still I do not know much about sealskins for all I have ever take is about a hundred in a season and I had plenty of salt [for] them."

He said, "Captain, ask old man Glass, the cooper. He is an old sealer and ought to know some about sealskins."

So I asked him what he thought of my taking twenty-two hundred skins. He said the same as Chipman; he thought it too many for the salt. I said to him, "How many do you think I ought [to] kill?"

He said, "Captain, I think you ought not to kill over fourteen hundred for you have got to come back and you can get them the next season."

I asked Mr. Gray what he thought about it. He said, "Kill what you think is right."

I told Chipman I would think it over and let him know in the morning before he went on shore. That night I divided [the] crew into three watches and I made short tacks so as to keep close in with [the] island and told the steward to have breakfast ready by four o'clock. The weather kept good all through the night with [a] light beating breeze. At four o'clock [I] had all hands called and by the time they had their breakfast I was close in with the island and I gave the order to clear away the two boats and told Mr. Chipman and [Mr.] Gray to go in and get what skins they could and stop as long as the weather was good and to tell Mr. Fuller to seal enough to make up seventeen hundred. I also told Mr. Chipman that I would go through the passage again and lay aback on the inside of the island and if I wanted them to come off I would set the colors.

At five o'clock the boats got away and I stood through the passage. After getting through, I hove aback first on one tack and then on the other. At nine o'clock it began to breeze up strong and the weather began to look threatening. I hoisted the colors for the boats' recall. At ten o'clock, they came on board with eighty skins. As soon [as] they took up the boats, I kept away for Norton's Harbor and had strong wind all of the way.

At three o'clock came to anchor in Norton's Harbor, went right [to] work and broke out one hundred and fifty barrels of oil, and landed it to make room for sealskins. The next day we was employed rolling the oil together on the beach and covering it over with earth to keep the sun from it and getting the salt on board, and at three o'clock in the afternoon I got underway. Got as far as Morgan's Bay, where I anchored for the night. In the morning I made an early start and I did not get no further [than] Breakwater Bay, the wind coming in northeast and lasting three days. On the fourth day the wind being light from the westward, I got underway and worked out to the seal rookery by ten o'clock, the two boats going where Mr. Fuller was. When the boats came off, they brought about seven hundred skins with them. After getting dinner, all three boats went in again to kill and skin what they could before dark, Mr. Fuller taking what salt he needed for about three hundred more skins in his boat. At dark when the boats [came off], two had two hundred more skins.

In the meantime the wind had dropped off to almost a calm, and the prospect was that we should have it thick. There was a

heavy bank to the northeast as I headed for Cape Français, and
about midnight it came in thick fog; but we did not lose run of the
land. As soon as it came good daylight I ran for Christmas Harbor, where I came to anchor about five o'clock in the morning.
After coming to anchor I gave orders to the steward to have breakfast at nine o'clock and to keep a lookout off the schooner and all
hands turned in until that time as all hands had been up all night.

After breakfast the crew went to work beaming skins and salt-
ing [them] down. At noon the wind hauled north northwest and it
began to blow heavy. When the wind first came, it was heavy
when it struck the schooner, that was, walking off with her
twenty-five hundred pound anchor and forty-five fathoms of
chain right out of the harbor. We gave her the big anchor and
sixty fathoms of chain. When that thirty-two hundred pound anchor got hold of the bottom, the schooner had to stop dragging.
Along in the afternoon it came on and rained, so that we stopped
work for the rest of the day.

For some four or five days we had bad weather. When the
wind came around light from the westward, I got underway and
went to the seal island and took Mr. Fuller off and the three
hundred skins that he had taken. After getting the skins on board,
I sent all three boats to look on the second big Cloudy Island. Before they got to the island it came in thick. All they got was ten or
twelve seal, but they said there was some three or four hundred
old whigs on the large island and the prospect looked good for a
large number of seal, but as it was thick they was unable to see. I
had to lay out all night as it was so thick that I lost run of the land,
and it kept thick until nine o'clock the next day when the fog lifted
and began to breeze up strong from northward. [I] did not try to
get into Christmas Harbor, but came to anchor in Breakwater
Bay, where we went to work and salted down the remainder of
the skins and getting in water and getting all things ready for the
passage to [the] Cape of Good Hope, we having on board eleven
hundred barrels of oil and about seventeen hundred sealskins.

The seal generally go in the worst places they can find and
they are very knowing both in the water and on the land. The old
whigs is the old male seal and clapmache is the female seal. In seal

rookeries the old whigs have their pods of females the same as the elephant, and the old whigs will fight to death for the female and even they will come for a man and seem to have no fear. In taking the seal we work around and get between them and the water and work careful to drive them together the same as a flock of sheep. After they get huddled together we let some of the men in and knock them down. The rest of the men keep on the outside to see that none of them get away, for if one old whig should make a break all the rest of them will follow. The seal club is about four feet long and made of hardwood and [is] thicker at one end, what a man can swing in one hand. One blow of the seal club on the nose will generally kill them, but I have seen men pound the head of an old whig to a jelly and they get up and go off into the water. As soon as we get through knocking down, we stick with a knife between the foreflippers to let the blood out of the body, then come skinning. First we cut just abaft the eyes, so as to leave the ears on the skin. Then cut right around the foreflippers close to the body and around the tail flippers, but leave the tail on the skin. Then cut a straight cut down the middle of the belly and in taking off the skin be careful not to cut no holes in the skin. After getting the skins off you beam them, that is, take off all of the blubber and flesh and let them soak in water [to] get the blood out of the skin then salt [them] down.

February the second the wind being southwest, I left Kerguelen Island for the Cape of Good Hope. We made the run in twenty-nine days, which is considered a good passage. We arrived at the cape the second day of April [*sic*, March]. We had one or two blows and some calms on the passage; the blows did not amount to much. The next day, after getting into Cape Town, I sent a telegram home, notifying my employer of my success. The second day after my arrival, I heard of two American barks that was ready to sail for the States and that they was going in ballast. I took a shore boat — it was blowing heavy from the southeast at the time — and I went on board of them both.

The first one did not want my oil so I went on board of the other, the *Moro Castle*, Captain Smith of Boston. I asked him if he wanted freight as I had about one thousand barrels of oil that I should like to get home to New London. He said that he should [like] to take [them] home for me if we could come to terms. He wanted to know what I was willing to pay. I told him that I would pay five cents per gallon. He said that he did not know much about oil and if I would make it the round sum [of] two thousand dollars he would take it direct to New London.[15] I told him that I would give it to him, and if he would go on shore with me we would have the charter made out and sign it at the same time, but there was one thing that he must do — that he must give me time to get my skins shipped before going to work on the oil. He said that he would give me four days and by that time he should be ready to take in the oil.

I went to work on the skins for shipping and I shipped them by the mail steamer to C. A. Lampson, London, England, and, at the expiration of the time set for taking in my oil, I hauled alongside of the *Moro Castle* and began discharging my oil. At the end of the fifth day we had the oil all out of the schooner then the cooper went [to] work setting up casks to fill with water for ballast for the schooner. The officers and men [were] working well, for I wanted to get away from Cape Town before the bark *Trinity* came in for I expected her in most any day. I did not want my crew to see any of her crew, for if they did I knew that [they] would get to blowing and would let out where I got my skins from and if they did tell I should have company at the seal rookery the next season.

In the meantime while getting ready I had four men run away, three foremast hands and one boatsteerer, Odell. One of the foremast [hands], I heard afterward, got drunk and he said that he did not want to run away but that he was shamed to come on board again. But Charles Odell I was determined to have again. I went and reported all four men to the consul and asked him if he would help me to get them back again, but he would have nothing [to] do with them.[16] So I took the matter into my own hands.

I gave out a reward for Charles Odell and Charles Fink and in a day or two I knew where they was and I could get them at any time. So about three days before I was ready to sail, I had Mr. Fuller and [Mr.] Gray come on shore after dark. I took two cabs and all three went [to] the house where they was stopping. I sent the man in to get them to come out. He was gone some time before he made his appearance with them. As they came out of the house, we all three nabbed them and put them in separate cabs: Mr. Fuller and [Mr.] Gray going in the one with Odell, and I took Fink in the cab with me, and we got them down to the boat all right. But Odell did not want to get into the boat until I went for him and I told him that if he did not get into the boat that I should pitch him in and that I should stand none of his nonsense. Into the boat he got and we took them on board and kept them in confinement until we left Cape Town. After getting on board I asked Charles Fink what he wanted to run away for. He said that Odell told him if they got home they could sell their information to Lawrence & Co. and then come out in one of Lawrence's vessels as pilot. I [told] Fink that I thought that he and Odell would make the trip to Kerguelen with me before they had the opportunity of giving information to Lawrence & Co.

About the same time the firm of Gready & Co. wanted to know if [I] wished to buy some good potatoes as they had some nice ones and that he could let me have them for one-third less than I could get them in Cape Town, potatoes being thirty shillings for one hundred and fifty pounds. I ask him where the potatoes was and where they came from. He said they was in an English bark and they came from Montevideo and the bark was laying in the docks and in a day or two the bark was going out into the bay. He gave me a sample and I took them on board and had them tried and they were fine potatoes. The next day I told him that I would take fifty barrels at ten shilling a barrel but he must get a permit from the custom house authorities to transship them aboard of the *Pilot's Bride*. He said that he would see that everything was all right.

About two days afterward that Mr. Gray came to me first at night and told me that the custom house officers had come off and seized the schooner and had put the broad arrow on the main mast.[17] I asked him what they had seized her for. He said for taking the potatoes on board without a permit. I told Mr. Gray that I could do nothing that night but I would see about it in the morning. The next day I went to the consul and stated the case to him.

He said he could not help me. I told him that all I wanted him to do was to state the case to the custom house authorities, but he would not even do that for me. At last I could stand it no longer. I told him that the American government ought to be proud of her representative at Cape Town and if Jim Blaine[18] had any more friends like him, he had better get the consulship at Cape Town.

In the meantime I had gone on board and asked Mr. Chipman how it was that the schooner was seized. He said that Gready's man came on board just at night and wanted a boat's crew to go on board of the English bark after sixty-five barrels of potatoes for the *Pilot's Bride* and said that Captain Fuller said the men could go if Mr. Gready would pay them for their trouble. Chipman said that he asked him if he had an order from Captain Fuller and he said yes but he had left it on shore. I told Mr. Chipman that he done wrong to let the men go and that he ought to have known that there was something wrong by Gready's man coming off after dark, but it is done now and I suppose that I shall have to pay a fine to get out of it. There is one thing — the potatoes are not paid for and I do not think will be this voyage.

Seeing that the consul would do nothing for me I went to my agent James Seawright & Co. and told them that the consul would not intercede for me and [asked them] if they could do anything with the authorities. They said that they could try and that they would see the collector and if they could do nothing with him then he would go and see the colonial secretary. But I must not think that it is going to be done in a day for you know government officials are slow to move; they will take their own time. That afternoon, Mr. Seawright went and saw the collector of customs and he said that he could do nothing but carry out the law, as the potatoes was found on board of the *Pilot's Bride* and they were taken on board without a permit and the crew helped to get them on board. That evening Mr. Seawright went to the colonial secretary and stated the case to him and he told Mr. Seawright that he would look into the case and let him know in a day or two.

About noon of the third day, the collector sent for Mr. Seawright and myself to come and see him at the custom house. We both went to [the] custom house. After greeting us, the collector said that he had got a letter from the secretary saying that he had looked into the matter of the *Pilot's Bride* seizure, and that he finds that Captain Fuller had no intention of doing wrong or going against the law, and that he is not wholly to blame in the matter but one of the citizens, Joseph Gready by name, and they might fine Captain Fuller a small sum and let him keep the potatoes on board. The collector said, "Captain Fuller, I shall fine you forty-five [shillings] and when you have paid the fine I shall dismiss the case with you and I want your deposition so that we can proceed against Gready and you are to keep the potatoes on board and not pay Gready for them."

[In] about an hour's time I was clear of the custom house and I thanked Mr. Seawright for what they had done for me, but he said he did not want no thanks but he was pleased to think that he could help me out of what looked like a bad scrape at one time.

A few days afterwards I went to the [consul's] offices and told him that I wanted my papers and when he was getting them ready he wanted to know if I was cleared from the customs house and [if] I had [my] crew all on board. I gave him my clearance from the custom house and it was no thanks to him for being able to say so but as for the men I had [the loss of] two [to] report and I must thank him for them getting away from me. After getting my papers and I was leaving the offices he bagan to wish me all the good luck there was in the world. I cut him short. I told him that I did not want any of his good or bad wishes. All that I wanted was to see a new and good man in his place when I got back from Kerguelen Island, and I got my wish for when I did arrive in Cape Town in 1883 I found a new consul in his place.

When I got my papers my intention was to get underway, but it came in thick and light rain, the wind dropped off to a calm. As I was going down the landing pier a man came to me and wanted to know if [I] would like to ship a man. If I did, he would like to ship with me. I asked him what his name was and how it was that he was stranded in Cape Town. He said that the vessel that he came to Cape Town in was condemned and that his name was John Reed and that he belonged in Boston. I told him that I did want a man but I wanted a man that had been sealing. He said that he had been sealing in the *L. P. Simmons* with Captain Potts at Cape Horn.[19] I told him that I would give him the one hundred and thirty lay and if he proved a good man I would do better than that when we returned to Cape Town, but that I could not ship him then as I was all cleared and if he wanted to go on them terms I would take him. He said that he was satisfied with the terms but he would like a little advance money to pay his board and to get a few things. I gave him some money and told him to be on board the schooner that night, and I sent Mr. Gray with him. Before they went I told Mr. Gray that I should not go out as long as it was thick but get the men on board that night.

April the second having been in the cape just one month to a day, the wind being fresh from the westward, I got underway and left Cape Town for Kerguelen Island, my old haunts. After getting clear of the land we had good weather and made the run in fifteen days. The fourth day after leaving the cape, I fell in with [a] school of sperm whale, it being almost calm at the time and the boats was lashed on deck and the whaling gear was in the hold. We got the two boats out and gear into them. The mate and myself went down. We both struck and [each] killed his whale, but in the meantime [it] began to breeze on harder and harder and I saw that we could not save them both so I let mine go adrift and all that we saved of the other was his body, it being [so] rough that I could not take in the head. We did not boil it out, not until we got to Kerguelen.

The day we got to Kerguelen Island, the wind was north and blowing strong. I ran close in with the Cloudy Islands, but found it too rough to make a landing. I ran into Breakwater [Bay] and came to anchor and went to work boiling out the sperm whale and it made forty-five barrels of oil. After taking care of the oil and good weather came on I got underway and went to the Cloudys and landed Mr. Fuller on the higher big Cloudy Island to take young seal, but not to trouble the seal on the island that had the seal rookery. After landing Mr. Fuller and boat crew I went looking for right whale but found none. I cruised about getting some elephant oil and two or three humpback whale.[20] Along in June, I told Mr. Chipman that I thought we had better take Mr. Fuller off the Cloudy and then go to Table Bay and pick up what winter elephant we could and try to fill our cask with oil before November, and he thought it a good plan.

The middle of June I took Mr. Fuller and his boat crew off. He had taken 397 seal. After taking them off I worked the coast along for sea elephant, and, when we got to Pot Harbor or Betsy Cove, the wind being to the westward, we had to beat in to the anchorage. As we opened out the harbor Mr. Fuller, who happened to be forward, said, "Captain, the *Trinity* has left lots of stuff on shore."

I took the glass and had a look and sure enough there was lots of cask and spare rigging on shore. After coming to anchor, the mate took his boat and we went on shore to find out to whom the things belonged and we found the things marked *Trinity* and on one cask marked "Captain Fuller." I had the cask opened and I found a number of boxes from my wife and some for the officers and a letter from Captain Williams to me. And the letter ran as near as I can remember after so many years has gone by to this effect, that he had arrived at Kerguelen Island the first of August and that he had tried to get into Three Island Harbor and was blown out of Royal Sound twice, so he had come to Betsy Cove and landed his spare stuff that he did not want to take to Heard Island with him. There was one thing that I remembered distinctly in the letter. It was: Captain Fuller, why cannot you come over to Heard Island before you go to the Cape of Good Hope and if the *Trinity* is lost you can take my crew to the cape and make one job of it, if there has anything happened to me. But I hope if you do come over you will find me all right. I gave the letter to Mr. Chipman to read. After he got through reading it he returned the letter and said, "Williams has got the damnedest cheek on him of any man that I ever came across, he wanting us to run out of our way five hundred miles to see if he is all right; well, I call that cheek."

After discussing the letter, Mr. Chipman wanted to know what I was going to do as it was almost a certain thing that the *Trinity* was lost at Heard Island. I told him that I did not know what to do about it as it was the last of July and Lawrence knew of my being in the cape the first of April and that I did not see the *Trinity* before leaving Kerguelen Island and Lawrence knows that if there had nothing happened that he would have heard from the *Trinity* by the middle of May at the last calculation. But I think by this time that Lawrence ought [to] have a vessel ready to sail for Williams's relief if they have not already sailed from home. But if there is no vessel out from home by October, I will land Mr. Fuller on the Cloudys and then I will go over to Heard Island and see if Williams is there, and if he is then we will take them off and take them to the cape with us. Mr. Chipman thought that was about the right thing to do.

So I kept on and went to Table Bay. We got all the elephant there was around Table Bay and about one hundred leopards, and then I went to Sprightly Bay and worked Cave Beach and we got a number of [elephant] from it and from Diana shore.[21] I then landed Mr. Fuller on Cave Beach with eight men and two boat-steerers. The two boatsteerers and three men went overland to Little Half-Moon and White Ash beaches to drive off the elephant that was on the beach. Mr. Fuller was to kill them as fast as they hauled up and Mr. Chipman and Mr. Gray was to get the blubber on board and to get all the elephant that hauled on Diana shore, for the season had got along to about the first of September and we only wanted a little more oil to fill our casks when the saddest accident happened on the voyage and cast a gloom over the whole vessel.

One day Mr. Chipman went around Diana shore and got a boatload of blubber and he got back about ten o'clock and he said that he never [saw] it any smoother then it was that day, and while we was getting dinner Mr. Gray wanted to know if he should take the boat and go to Cave Beach and get what blubber Mr. Fuller might have, as he could land most anywhere in a tub as he expressed it. I told him that he might as it was calm and if Mr. Fuller did not have a boatload to go as far as Boat Bay and see what was there. Right after dinner he went and the weather keep good until about four o'clock, when it began to breeze up, and by five o'clock it was blowing [a] heavy gale and the bay was feather white, the wind coming down off the mountains on the west side of the bay in heavy puffs, and when they would strike the vessel she would heel over so that her quarter boat would almost touch the water. When we was getting supper, Mr. Chipman said that he thought that Mr. Gray would have to stop with Mr. Fuller for the night. I said yes for he would be [a] very foolish man to try to get on board in such a blow as this and we thought nothing more about him. Through the night it blowed heavy, but along toward morning it began to moderate and at sunrise it was light wind.

After breakfast one of the men came aft and said that he could see smoke on Cave Beach, he believed. I took the glass and looked and could see that some one was making a big smoke. I told Mr. Chipman that I thought Mr. Gray had lost his boat in the last night's blow and he had better take his boat and go down and see what the matter was. He started right off and was gone perhaps one hour when I [saw] him start to come on board, and, when the boat got close enough to see good, all I could see was two extra men, and I began to fear that something had happened to Mr. Gray. When the boat was coming around the stern I asked Mr. Chipman what had happened. He said that Mr. Gray's boat had capsized and Gray and three men was drowned and that he had the bodies of Mr. Gray and one of the men; the other two was lost with the boat.

After getting the bodies out of the boat, I asked Mr. Chipman how it happened. He said what he could understand from Mr. Fuller it was all through carelessness in trying to come through a passage [so] that he might cut off a few yards in distance, and Mr. Fuller said that Gray came down to Cave Beach and landed, leaving his boat in the water, it being so smooth, and said, "Well, Mr. Fuller, I have brought you all the pies and cakes that the steward has baked, and how much blubber have you got?" Mr. Fuller told him that he had only one bull elephant that would make about a third of a boatload.

"Well," he said, "Mr. Fuller, I am going down to Boat Bay and perhaps I will get one there and I will stop coming back and get what you have." Mr. Fuller told him that the blubber was down on the lower end of the beach and that when he saw him coming from Boat Bay he would have the blubber down to the water. Mr.

Gray got into his boat and started. He and the boat crew seemed to be in the best of spirits to think they had nearly blubber enough to fill all the casks with oil.

Mr. Fuller said soon after Gray left, he and the men went on the hill and see Gray go into Boat Bay and when he came out of the bay, then he and the three men went down the beach where the blubber was and had just got the blubber out of the hole when he happened to look off shore and see Mr. Gray's boat coming through the small passage between [the] mainland and [a] small rock. As the boat was shooting through the passage, the head of the boat grounded and over she went, and, instead [of] Gray getting on the rock or mainland as he ought, he stopped by the boat. Mr. Fuller did try [to] have them to leave the boat while they were in the passage, and he [himself attempted] to get across the big water stream that runs out at that end of the beach and go to their assistance and help them to get on shore, but the water was too deep and running so swift that they could not stand on their feet.

By the time they gave up trying to get across the water stream, Mr. Gray had got the boat rolled over right side up and one man got in her and was bailing out the boat when some of the crew tried to get in and over she went again. By that time the boat had drifted out of the passage and Mr. Fuller called out to Mr. Gray not to try to get in the boat but to hold on to her and he would go and get his boat and come to him as quick as he could, his boat being about three-quarters of a mile away at the upper end of the beach. Mr. Gray waved his hand to let Mr. Fuller know that he understood what he said. Up the beach the four went and Fuller said that he never traveled three-quarters of [a] mile as quick as he did that [day], and, the three men was close to him all the way, they knowing that six lives hung upon their exertions and that they must put forth their whole strength to rescue them from their peril. When they got to the boat they found that she was frozen down and it took some time [to] loosen the boat.

After getting [the] boat loosened came the trip to get her down to the water and Mr. Fuller said it was a hard pull for them but they knew that they must do it and that it must be done quickly so every man pulled his pound and down to the water she went. He

said that he never stopped for nothing. Into the boat they got and
he believed that the boat never went so fast with a whole boat's
crew as she did with them three men. When he got close enough
to see, he could see five men on the boat and he saw one man let go
of the boat and disappear. When he got close to the boat another
man let go of the boat and was going down when one of the men
had just time to reach him with the boathook. That left three men
with Mr. Gray and they kept to the boat until they was taken off,
Mr. Gray being the last one. After getting him in the boat the
only words that he spoke was, "Mr. Fuller, you have come at
last," and then he dropped off into a drowsy state and he neither
spoke or moved from the time they took him off the boat.

Mr. Fuller said that it began to breeze up soon after taking
them off the boat and he started for the schooner and when he got
half-way the wind came so strong that with only three men he
could make no headway. At last he had to give up and go back to
Cave Beach. As soon as he landed, he took Mr. Gray and the man
in the cave and went to work on them, the other two men being
able to take care of themselves. He worked on Gray and the man
all night but to no effect. He thought that if it had not blowed so
heavy and he could have got on board that we could have saved
them both, for [they] took so long before they could get to work
on them, but he tried all the remedies that he could think of but
nothing that they could do would fetch them around, so at daylight they made a smoke to let us know that something had happened.

The next day we buried Mr. Gray and the [other] man, and
one of the men cut the names of all four on a board and we put it
up as a headboard to the graves. After performing the sad and sol-
emn burial service for Mr. Gray and the three men, I did not stop
long in Sprightly Bay. I told Mr. Chipman to go down to Cave
Beach and tell Mr. Fuller to get the men from Little Half-Moon
and White-Ash beaches and come on board. Two days [later] Mr.
Fuller came on board and he said to me, "Captain, I am glad to get
away from Cave Beach."

I asked him why. He said everytime that he went on the beach
he imagined that he could see and hear Mr. Gray and the men hol-

lering and struggling in [the] water. He [said] it was awful to see them drown before your eyes and you not able to do nothing for them.

As soon as Mr. Fuller came on board, I got underway and went to Table Bay and got what elephant there was and boiled out all of the blubber and I found that I lacked about one hundred barrels of oil to fill my casks. I told the officers that I should not stop at Table Bay no longer but should work along down and go up Royal Sound to Three Island Harbor so as to get there by the first of October to see if Lawrence [and Co.] have sent a vessel to look after Williams and if he has not sent a vessel then I must land Mr. Fuller on Cloudy Island with a boat's crew and then go over to Heard Island, for we could get all the oil wanted while waiting for Mr. Fuller to take the seal on Cloudy Islands.

After taking all the elephant around Table Bay, I got underway and worked Greenland Bay and the little beaches along [the way]. I came to anchor in Three Island Harbor and I stopped there until the last of September. Finding no one from home we boiled out the blubber that we got coming along down. The first day of October we left Three Island Harbor and that night came to anchor in Shoal Water Bay. The next morning after breakfast I told the mate to get underway, the wind being west northwest and light squalls of snow. When we got off Cape Digby, which was about noon, the wind dropped off almost to calm, what wind there was being from north northwest and [with a] heavy short head sea and the tide being ahead we made slow work of getting along.

About five o'clock the wind hauled more to the eastward so that we could lay along the land and came in hazy, and at sunset we was off Mount Campbell Point heading in for Betsy Cove. It began to thicken and snow and I asked the officers [whether] they thought we could keep run of the land and get into Betsy Cove all right. At six o'clock the wind freshened up strong and came in thick snow storm and we lost run of the land. I told Mr. Chipman that I thought the best thing we could do was to haul off-shore, and, after getting clear of the Rocks of Despair, heave to for the night. He said that he thought it better than running in on the land when we could not see where we was going and wanted to know if we could fetch clear of the Rocks of Despair. I told him that I thought we could, as the wind was well to the eastward. We came about and stood off-shore, heading north northeast by compass and by that time it became so thick and dark that we could not see more than half the length of the schooner ahead. I put the two boatsteerers on the lookout and told them that if they saw kelp to let me know, and we stood along about three quarters of an hour.

I began to think that I was all clear of the reef, when the men on the lookout sang out, "Kelp ahead!" I told the man at [the] wheel who happened to be Glass, the cooper, to put the wheel down and sang out, "Stand by for stays!" The schooner came up in the wind and there she hung for about two minutes and then fell away on the same tack again. I knew that there was not room for her to fill away and gather headway so as to try and tack again before she would be on the reef. I told Glass to put the wheel hard port and at the time told them to lower the peak of the mainsail and to stand by to wear ship. By the time that I had gave the orders the schooner was falling off fast, when Mr. Fuller who was standing on the lee side sang out, "Hard down, there is a rock on the lee bow!"

I sang out to Glass [to] put the wheel hard down and sang to the men to hoist the main peak and I jumped to the lee side and caught a glimpse of the rock, which was not off more than half [the] schooner's length, and the schooner was coming up fast into the wind. But it was of no use; she could not get clear of the rock. The next minute a sea came and picked her stern up and landed it on top of the rock and as fast as the sea would throw her stern up she would forge ahead and draw off again. [The] third sea carried away the rudder. The schooner keep forging ahead and at last got clear and as soon as she got clear we took in the sails and came to anchor in twelve fathoms of water, it being so thick that we could [not] see where we was and I was afraid that she would get stern way on and drift down on the reef and I did not know what was ahead of us. I gave her sixty fathoms of chain and the 2,200-pound anchor and we cleared up the decks, furled the sails, and Glass came and said that he thought the rudder was all right as he could put it hard up and hard down. While they was furling the sails I had some of the men pump her out but found that the schooner was making no water. But it still kept thick and snowing. I told the men to get their suppers as we could do nothing as long as it was thick and dark.

After getting supper, we let go the big anchor and paid out about fifty fathoms of chain so if the wind should haul to the westward she would lay in the position. After doing all we could for the safety of the schooner, we divided the crew in three watches. I told them to let me know when it stopped snowing or when they could see off any distance. I then went down into [my] cabin to see if [I] could get a little sleep, but I found that it was impossible for me to get to sleep. My mind would wander to the situation of the schooner and at last I gave it up and went on deck with the anxiety to know where we was. I could not stop in no one place; I had to keep moving about first on deck then down in the cabin. Oh, how anxious I was for daylight, for the night was long; it seemed as if day never would come.

At last about two o'clock they came and called me and said the wind was about the same but it had stopped snowing, and they could see that we was close to the reef. I went on deck. The wind was still to the northward and blowing strong [with] a heavy sea, the schooner pitching heavy, taking water over the bows, but still she did not seem to be riding heavy to her anchors. I could make out that we was close to the Rocks [of] Despair and that we was laying in about the middle of the reef, the reef forming a small bight or horseshoe bend. As my eyes got accustomed to the darkness, I could see through the gloom breakers on the port beam running around the stern and extending out ahead on the starboard bow, and in the darkness it was an awful sight to look at: the reef one sheet of white foam with points of black rock sticking out of the water.

After seeing how we was situated, I went in the cabin and called the steward and told him to have breakfast by daylight, and that would be about six o'clock at that time of the year, as we must get out of the place before the wind hauled to the westward if it

was possible for us to do so. Soon as day began to break the steward called all hands and we got our breakfast. While getting breakfast, I told Mr. Chipman to take up the big anchor but before he
broke it away to run out a kedge to the westward for that was the
way [we] must get out. Soon as we got through breakfast we all
went on deck and Mr. Fuller was looking over the stern to see if
the rudder was all right and he came to me and said that [the] rudder was gone and nothing left but the rudder head. Here was a
misfortune that we did not look for and we must have a rudder of
some kind. The mate wanted [to] know what we should do now
[without] a rudder. I told him that I did not know, unless we
worked the schooner with her sails. One of the men came to me
and said, "Captain, take the main boom and make a rudder out of
that using it on the same principle that you use a steering oar in a
whale boat." He said that he was in a ship that lost her rudder and
the captain took the main yard and used that until they arrived in
Cape Town. I told him that it was a good idea.

We went right to work unrigging the main boom, which took
some time to unrig, and getting it in place and, while we was
doing it, the wind hauled to the westward and commenced to
breeze up strong, bringing short heavy [seas] with it. After getting
the main boom where we wanted it and getting the trysail bent to
take the place of the mainsail, I asked the officers what they
thought about getting underway. They thought it was a hard look
just then and that we had better hold on until it moderated some
for if we let go of our chain we would drift down on the reef
astern, which was off about twenty-five yards. I waited until
noon, hoping that it would moderate, but it was hope against
hope; instead of moderating it blew the harder and [with] more
sea.

At four o'clock the weather began to look worse and the
barometer began to fall fast, and there was a heavy bank of clouds
in the northwest and extending around to the northeast. I got the
officers in the cabin and asked them what they thought of the situation and what we had better do to get out of it. They said they
did not know what advice to give, but they thought we ought to
do something before night. I then told them what I thought about

it. I told them as for getting the schooner out from where she was, it was impossible with the wind and sea against us as it then was blowing, but I thought she would hold on where she was with the anchors and chain that the *Pilot's Bride* had. And to make sure of our safety, I thought we had [better] get on shore in Betsy Cove before dark and if the schooner did ride it out, all right. After it moderated we could come off to [her] and get her out and, if she did not, we would be in safety but would be without a home, and if any of them had any better plan let us know it. But they thought that the most feasible plan and they thought the vessel would ride it out all right. I told Mr. Chipman to tell the men what we was going to do and to get a few things such as blankets, but not too much, and to be quick about it for we have a long way to pull before dark and to see that the chains was well stoppered, and I would get a few matches and see that things was all right in the cabin. I took two blankets, two pair of stockings, one pair [of] drawers, and [a] long oil coat and a case of matches. After seeing that [the] fire was out and [there were] no lights in the cabin, I came on deck and locked the door and I found [the] crew all ready to leave, Mr. Chipman and Mr. Fuller going in their own boat; Glass, the cooper, taking charge of the starboard quarter boat; Mr. Fuller being the last to get away, he stopping to haul the boats' falls taut and make them fast.

It was four thirty when the boats left, and at dusk we landed in Betsy Cove both wet and cold for long before we got in. It was blowing [a] strong gale from [the] northwest and we shipped a good number of seas coming in, the distance being about eight miles. After getting the boats up on the beach and making them fast, we went to work making a shelter for the night. We took some of the *Trinity*'s spars and sails and made a large tent and then we made a large fire in the middle, for when we landed we was fortunate enough to get a sea leopard and we made fire wood out of the blubber. About the time the tent was finished it began to rain, the wind hauling to northward, and that night it blowed heavy and rained. I do not know where the most water was, inside of the tent or outside, for it came through the canvas in streams and it was impossible to keep dry. We spent a very disagreeable night and was glad when morning came. But it still keep blowing and raining that day and night, but we went to work [and] made the tent a little more habitable and cut a ditch around the tent to let the water run off, and the second night was [a] little more agreeable than the first night. The third day the wind moderated and came in thick fog. In the meantime some of us was continually going up on the high land to see if they could see the schooner, but it kept so thick that we could see no distance. But I told the officers that I had no hopes of the schooner riding out such a gale of wind and sea as this was.

On [the] morning of the fourth day I was up bright and early as the wind had come around to the westward and went up [to] the top of the land and had a look off to the Rocks [of] Despair. I could see the rocks but no schooner. She had disappeared. It was with a sad heart that I looked off to the rocks thinking of the loved ones at home who was expecting us home in seven months' time: when shall we meet again? Shall we ever? But I was thankful to my God for all of His great blessing to me for giving me wisdom to leave the schooner when I did. If we had stopped by the vessel, not a man would have been saved. Only my God knows with what a sad and sorrowful heart that I returned to the tent knowing that I had twenty-two men to look after and some of them hard characters; one or two of them had not been out of prison long before we sailed from home and one of the boatsteerers was not much better, but there was some good men among them. But in going down I made up my mind that there must be a head and that I should be that head and if any of them did not comply with my orders then some one should get hurt. I knew that my officers would stand by me and some of the men.

I went in the tent and told them that the schooner was gone and in all likelihood we would have to stop on the island about a year as they will not know or mistrust anything has happened to the *Pilot's Bride* until March, and I did not expect no one until [the] last of September or the first of October. They [should] try and live [at] peace and do as near right as they could; we would share alike in whatever we had, both in work and in provisions that we had; and I though we had provisions enough to last one year by being economical. We would use the *Trinity*'s first and make it spin as far as it [would] go. After getting something to eat I said, "Men, the first thing that we must do is to make this shanty habitable for we must stop at Betsy Cove as long as we can get food [to] eat. Now men I want one to go with Odell down to Black Point this side of Mount Campbell and while you are gone the rest of us will get to work on the shanty and see what we can do."

As soon as Odell and Fink had [left], we went to work on the shanty. In the first place we levelled off a space about thirty-five feet long and about twenty in width. Then we built a wall about three feet in height, of sod on three sides and a ledge forming one end. On top of this wall we laid two of the *Trinity*'s spars, one on each side, and took one of her topmasts for [the] ridge pole. After getting the spars in their place, we took some of the *Trinity*'s running rigging and made lattice work of it by making one end fast to the spar that laid on the wall and letting the rest run over the ridge pole and under the other spar back and forth the whole length of the spars and then hauling them tight like the strings of a violin. Then we took two of the *Trinity*'s sails, the foresail and topsail, and spread them over the ridge pole and letting the ends come down on the ground. After smoothing the sails out and lapping the edge over one on top of the other so as to shed off the water, we put sod of earth on the sail to keep the wind from getting under the sail and to keep the sail taut over [the] ridge pole.

After we had been to work [a] little while the men came to me and said, "Captain, we do not want you to work. You do the bossing and tell us what you want done and we will see that it is done."

I told them that I could not stand and see others work and I do nothing, but as long as I was able I should try and do my share of the work. While [we] was at work on the house two or three of the men was pulling grass and spreading it out on the rocks to dry in the sun. About four o'clock we had the shanty very comfortable and then we got in the grass for our beds and over the grass we spread an old sail. When we turned in at night we made a field bed and we turned in all together for the warmth.

About dark Odell and Fink returned. We was just getting

supper. I asked Odell what he saw. He said all along the coast as far down as Black Point the shore was strewed with the *Pilot's Bride's* wreckage and there was [a] large number of full casks of oil and casks with their heads knocked out and some of the full casks was thrown right up in the grass. He said, "Your washstand has come on shore and there is not a scratch on it. The door was locked so that we did not see what was in it. We [didn't] see any provisions, but we did see some clothing which we put up in the grass to dry. But Captain Fuller, the best of all: the whiskey barrel has come on shore all right and I rolled it up in the grass and before we came away we took a good swig out of it." I told him that I could see that without he telling me of it.

That night we had the fire in the middle of the shanty and as we was sitting around the fire I told the officers and men that two boats must go to Norton's Harbor and get a cask of flour and to see if they could find a ship's copper [pot] that used to be there the first good day. And tomorrow we must get to work and see if we can build a chimney. Who is architect enough to build one? Ed Carroll spoke up and said, "Let old man Glass [do it] and we will get the stone and mortar. Up on the hill is some red sticky clay, just the thing for mortar." I told them that Glass should have the bossing of the job and that I would go with three or four men and get some blue night hawks.[22]

The next morning the men went to work on the chimney. Mr. Chipman, myself, and three men went nighthawking. Fortunately for us, there happened to be in each boat a garden spade and one shovel and they came in play in digging out birds and rabbits as both the rabbit and night hawk burrow in the ground. That afternoon we came back to the shanty with about three hundred night hawks and as the men had the chimney done we all went to work cleaning them and as fast as we cleaned the steward and cook went [to] work frying them, we having two large cast-iron frying pans. That night we managed to get away with about two hundred of them. I told the steward that he had better have the rest of them in the morning.

The next day having nothing to do most of the men went to Black Point. They said they wanted to see what they could find in clothing, but I knew it [was] the whiskey barrel that they wanted to find. At dark they did not get back — only four of the Portuguese, and some of them did [not] get back until late that night. Come to find out from the Portuguese they had got into the whiskey barrel and had got full before they started to return and when they was coming back when crossing the big water stream, one of the men, Tom Flaherty, fell down in it and could not get up he was so drunk. [He] would have drowned [if not] for one of the men who happened to be [a] little more sober than the rest. After they got him out [of the] water stream, they left him to get along the best he could, so they all had come to the shanty but Tom. Mr. Fuller went after him and got him home or he would have perished through the night.

I did not say anything to them but the next morning I told them that I was going down to Black Point. Three of the Portuguese wanted [to] go down with me and Edw. Carroll and John Reed. I told them all right and I told Carroll to take the boat hatchet with him. When we had crossed the big water stream, we began to fall in with the wreckage. At one place was the top of the [deck] house and it was whole and on top of it was two linetubs full of whale line. Carroll wanted to know how I thought it was that the house should come on shore with the two tubs on it and the whale lines did not get spilled out. I told him that was easy to account for. I asked him if he did not see all the empty casks that was laying around with their heads out. He said he did. "Well, Carroll, all of them casks was full of oil and when them casks got on shore and had their heads knocked out all that oil was in the water. There was no top sea with all oil on the water, nothing but a heavy running swell, and I suppose the top of the house got on top of one of them heavy rising swells and landed just where it is now and not another came up to [it] afterwards."

A little farther along we came to my washstand. I knocked open the door; I found a pair [of] shoes, one slipper, a chamber vessel full of soap, and a two-gallon demijohn, empty. I told them [to] take the demijohn along with them. After looking at the different things along, we came to the whiskey barrel. We filled the demijohn with whiskey and [a] bucket that we found alongside of the barrel. After filling them we rolled the barrel over and let the remainder run out and I stood over it until it was all out. As the whiskey was running out the men were washing their hands and face in the whiskey and the men thought that I was doing a good thing. I told the men that they might take the bucket to the shanty and give them all a drink around, but the demijohn I should put away in case we should need spirits.

As we returned [to] the shanty the men picked up a number of different articles of clothing and we came across a small cask and it was full of something. We called it flour as it was marked flour on the head and it was heavy. I told [them] that we had better let [it] alone as it was a long way from the water and some day we would come down and open the cask. After getting to the shanty I called Mr. Fuller and told him to take the demijohn and bury it somewhere and to let no one know of it, so that we could have some spirits in case of need. Before we got to the shanty I requested the men to say nothing about what we had done with the whiskey but let them find out themselves.

We had been on shore about ten days when the wind came around light from the eastward. I told Mr. Chipman and Fuller they had better go to Norton's Harbor and leave a note and get the flour and things that we wanted and fetch down a box of soap as there was two boxes up there and after they get up to Green Point they would have fair wind. They had made two boatsails out of the *Trinity*'s royals while they had been waiting and they got away early as they had about forty miles to go. Soon after [the] boats left, one of the Portuguese said there was [a] large piece of wreck up the bay. We all went up where it was and we left the steward at the shanty to get dinner by the time we got back. When we got to the place we went to work and knocked it to pieces and got it down to the house and that same day we got two elephants and leopard close to the shanty that had hauled up on the beach. After we went to the wreck, after killing and taking the blubber off we dug a hole in the ground where the water would run in the hole then we put the blubber in the hole, the water keeping the blubber fresh a long time.

The way we make a fire out of blubber is this way. First we

take a little wood and start a fire. After getting it a'going, we cut up the blubber in small strips and mince it fine and then lay the strips on the wood fire. After the oil gets running the fire will burn nothing but the blubber. We only use wood to make coals and the blubber makes a very hot fire. The only drawback with blubber is the smoke and soot that it makes at night. When the steward and cook gets through with their work at night they look like negro minstrels. They did their work cheerfully; once in a while they would have a growl and afterwards they would be the best of friends.

The next day the weather being fine we took the boat and went down to Black Point and got [a] boatload of wood and all the clothing that we could find and got back to the shanty about noon and about three hours afterward the boats came back from Norton's Harbor with the flour, copper [pot], and one box of canned goods, and [a] box of soap. They also brought down an iron coal-tar barrel. We put the tar on the canvas over the shanty, then we cut the barrel in two and made two coppers out of it. The canned goods tins we used for dishes to eat out of and we made wooden spoons. As soon as the boats got back the steward took charge of the cooper and soon had a rabbit stew underway and the savory smell from it set us all a'longing for it to get done. For the last ten or fourteen days we had been living on fried meats, and when the stew [was] cooked and dished up I [do] not know if ever I relished a meal as I did that [night]. From that time out we had a rabbit stew once a day until we was taken off the island.

That night the mate said that he had put the note on a pole and made it fast and stuck the pole in the ground. If there was a vessel went into Norton's Harbor, the [vessel] could not help seeing it. I told him that I did not think it likely anyone would go there before we did. And the only place they would be likely to go to would be Three Island Harbor and I thought that a boat ought to try and go there and leave a note. "But, captain, how about getting back? I know that the boat [can] go there for the wind will be fair going but it's the getting back is what I am looking at."

I said, "You can get back easy enough for after you have been to Three Island Harbor you pull over to Malloy Point and haul your boat up on the beach and then walk back overland. I do not think it will be over forty miles to walk."

"Well, captain, I think forty miles enough to walk in one day but how far do you call it by water?"

I said, "Well[I] think it can not be far from one hundred and fifty the way the boat will have to go."

"Well, Captain Fuller, it will be risky business for one boat to go alone."

"Yes," I said, "But we can spare only one boat. The other boats we must keep in case of accidents. I think it can be done and we ought not to let no chance slip by without trying to do some thing for our relief for there may be[a] vessel call at Royal Sound when they would stop at no other place and we must get there somehow and leave a note. If no one will go, I shall go myself if I can get a boat's crew that will volunteer to go with me."

The men spoke up and said, "Captain, when you want a boat's crew we are all willing to go and do what you want;[if] you want to try and get[to] the Cape of Good Hope, we will[be] willing to go with you." I thanked the men for their good will and told them that we should not go before the middle of November.

The next day Mr. Chipman wanted to know if he should not [take] a boat and go to Black Point and get some wood and see if they could find some more clothing as it was smooth and fine weather; the boat crew wanted to go. I said yes, by all means go; they may as well be going there as doing nothing. After they had been gone sometime one of the men said, "Captain Fuller, all Chipman and the boat's crew want down to Black Point is the whiskey that is down there." I thought it best not say what I thought, but said I thought they would have a good day down there. About noon the boat returned with some wood and cloth-ing. After[they got the] wood up and hauled the boat well up on the beach and made her fast, Chipman and myself was walking toward the house. I said in a casual manner, "Well, Mr. Chipman, the men did not find what they went after, did they?"

"No, and I am damn glad they did not find the whiskey and you did a good thing to let it run out."

That afternoon I was down on the point at the mouth of the

harbor where there is a cave where I usually spent my spare time in walking up and down as the cave was in the lee and no wind could blow in no matter how the wind blowed. Odell, one of the boatsteerers, came down and we got to talking about Black Point. After [a] while Odell said, "Well, captain, when we was going over there this morning old man Chipman said, 'Well, boys we will have one good drink of whiskey but you must not drink too much and full you know.' After we had landed and hauled the boat up on the beach [and took] the boat bucket to draw out the whiskey, then all of us started off on the run for the whiskey barrel and we could all outrun Chipman, so he hollered for me and the rest to stop running but we keep running and let him holler. He kept saying, 'Boys, let's all go together; there is whiskey enough for us all,' but we did not stop until we got to the whiskey barrel. By the time I had taken out the bung, old man Chipman [came] along and I took hold of the barrel to fill the bucket. I found the barrel was empty and Chipmen said to me, 'Odell, why don't you fill the bucket?' I told him, 'Because there is nothing in the barrel; it is empty.' Chipman looked at the barrel and then said, 'Well, boys, damn his whiskey. I do not want any of his damn infernal stuff.' And we told him that it was a very good thing why Jack did not eat his supper for there was nothing for poor Jack to eat. And Captain Fuller you would have split your sides laughing to see the old [man] running and trying to keep up with us and the look he gave at the barrel when he found it empty."

"Well, Odell," I said, "I can imagine [it] all without being there and I knew what they went after before they started to go to Black Point." Then I said to Odell, "You know, Odell, as well as I that the first thing Mr. Chipman would have done was to get drunk and then he would get to abusing the men and there would have been a fight and someone would either get hurt or killed and I am thankful that I let it run out and that you did not find none in the barrel. You know it was only a few days ago that you and most of the crew went down there and you all got full and when you and the rest of the men was coming back Thomas Flaherty fell down in the water stream and would have drowned if Carroll had not helped him up and out of the water stream. Then you all left him to get home the best he could and if Mr. Fuller had not gone out to look for him he would have stopped out all night and very likely would have frozen to death. Odell, you know when whiskey is in, senses is out."

"Captain," said Odell, "Flaherty could get along as well as we could. We did not make him or tell him to get drunk."

"I know very well you did not make him drunk, Odell, but you might [have] looked after him." But I might as well have talked to the wind for Odell belonged to the class of young men that think it a great thing to get drunk whenever they get the opportunity of doing so. He only laughed at what I said to him.

From time to time the men kept going to Black Point, picking up clothing and blankets but they was full of oil and it was sometime before we got the oil out. When we did get the oil out the clothing and blankets came handy and they was as good as new to us. The men was very good. Whenever they found any clothing or thing that belonged to any one they would give them up. One day four of the men went down to Black Point and they did not get back until about midnight. All the rest of us had turned in for the night, thinking they was going to stop at the point that night. The next morning when I was away from the house Carroll came to me and said, "Captain Fuller, you know that cask that you thought was flour?"

"Yes," I said, "I remember it well."

"We opened it yesterday and found that it was full of sealskins and they are in good order."

"Well, Carroll, I hope you put the head in the cask again for they must stop where they are. We can not bring them up to Pot Harbor for we have no salt to take care of them."

"Yes, captain, we put the head in and the skins is first class and there is plenty of salt in the skins."

"Well, Carroll, how came you to take the head out of the cask? If it had been flour we should have been out so much eatable food and a cask of flour would have lasted a long time made into slapjacks and duff[23] for you know that when the head is out of a cask of flour and the weather gets at it, more especially if it had rained, it would have spoiled."

"You see, captain, it was this way. We came to take the head out. You know the cask laid bung down. We got a big lever under it and rolled it bung up and knocked the bung out and found it sealskins instead of flour. Then we thought we would look and see if they was all right. Now, Captain Fuller, to whom does the sealskins belong to? Do not they belong to the ones that finds them?"

"Carroll, I rightly do not know, but I think they belong to the underwriters. You know that everything belonging to [a] vessel that is of value or cargo that is saved from her goes to the underwriters; that is my impression."

"Captain Fuller, I think they ought to belong to them that found them."

"Well, Carroll, no matter to whom the skins belong to I will take charge of them for [the] present and when we get where we can find out I will give the skins up to them that the law compels me to. You can make up your mind. I will do what is right about the skins, and you can tell the rest of the men what I say about the skins."

"Yes, captain," said Carroll; "I should like to ask you one thing — that is, if the skins do not belong to the underwriters do all the crew come in for a share in the sealskins?"

"Yes, Carroll, I do think they ought to have their share of them."

"Well, captain, Cape Town Jack [Reed] and Odell do not think they ought to have a share. They say you and the five men are entitled to a share of the skins."

"Well, Carroll, if Odell thinks that way, how is he entitled to a share? He was not with us, you know, when we found the cask."

"He was with us yesterday, you know, and that makes the five men."

"It will not make no difference to me what Cape Town and Odell think; I will take charge of the skins and as all the crew helped to get the skins they are entitled to their share of them.

Take it home to yourself, Carroll; would you like it if you had not been one of the men that found the cask of sealskins, and one of the men [who] did find them should say to you, 'Carroll, you are not entitled to the skins no matter if you did help to get them — you did not find them,' I think you would think it hard lines."

"Captain, I do not know but you are about right, but I do think that some of the crew ought not to have a share in the skins. They are so (d——d) lazy that they hardly leave the house."

"Carroll, you know we must do good for evil and they are a good number of years older than you and me and they cannot walk the same as younger men you know."

"Yes they can, captain, if there is whiskey to walk for."

Some days after the finding of sealskins, one night after supper we got to talking about the chances of getting away, when I had a chance of sounding [the] officers and crew of an expedition that I had long contemplated: someone taking a boat and going to Three Island Harbor and leaving a letter. I was in hopes that one of the officers would make the offer to go on the trip. It came about by one of the men, Melrose, wanting to know if a vessel would not go into Three Island Harbor sooner that Pot Harbor if a vessel should come from home looking for Captain Williams at Heard Island.

"Yes," I said, "Melrose, they are more likely to go into Three Island Harbor than any other place on the island, for they usually make Three Island Harbor their headquarters when working Heard Island for years. Now I think someone ought to take a boat and go to Three Island Harbor and leave a letter in the house on Hog Island called Fort Independence."

"Yes," said Mr. Chipman, "But, captain, it is a dangerous undertaking for one boat to go alone."

"Yes," I said, "I know it has its dangers but I think it can be done by working careful and taking good days and do not be too hasty to get along. The dangerous part will be Cape Digby. After getting around that point you will have good beaches to land on most of the way to Royal Sound."

"Well, captain," said Mr. Chipman, "You can excuse me; I think I shall stop at Pot Harbor rather than take the trip."

I said I did not ask him or any one to go. Whoever went must voluntarily do so for I did not ask no one to go where I did not wish to go myself. But I still [thought] some of us ought to go.

After I had got done speaking three of the men spoke up and said, "Captain, if you will go we are willing to try it with you." Then another said, "You can count me in, captain," "And me too," said another. So it ran around among the men until they all had volunteered to go if I went. I thanked the men for their offer and told them that we should not go until the middle of November, but they could be getting ready for the trip.

Then I told the men that I did not have no writing paper and asked them what I should use to write the letter on. Then Carroll, Melrose, Odell, [and] Cape Town Jack said, "Captain, the steward has some writing paper and we will make him give you all you want. If he will not give it to you we know where he keeps it and we will take it by force." The next morning after breakfast the men called the steward out of the house. When they had him outside they all got around him [and] Carroll said, "Steward, we want some writing paper."

The steward said, "Carroll, you cannot have a damn sheet of my paper."

"Look here, steward, you say you will not give none of your paper to us; well, give it to the captain. What is the good of your paper to you I should like to know?"

"Well you can not have it, Carroll, for I want to keep a journal myself."

"Well, steward," said Carroll, "if you do not give Captain Fuller all the paper that he wants we will take it all from you for it is as much for your benefit as it is for ours."

"Well," said the steward, "I think it mighty hard that a man cannot do what he likes with his things."

Then they all had some thing to say which was not very complimentary to him. However, the next day the steward came to me and gave me two sheets of paper and offered to give me more if I wanted it but I told him that was sufficient for my use. But that did not save his paper or his journal that he had begun to keep, for sometime after he gave me the two sheets of paper I wanted some to make a small journal so I asked him for some. He said, "Yes, captain, you can have all you want," so he went to his bag to get me some but he could find no paper. He turned around to me and said, "Them damn cursers has stoled every sheet of paper that I had!"

I was in no doubt but I knew the parties that had taken the paper so one day I asked one of the men. He said yes he was one of the parties that had taken the paper. I asked him why he had done it. He said, "Captain, you ought to have read the journal he was keeping. He was giving everyone the devil and my word giving old man Chipman the devil's own raking down and we did not [want] him to be writing a damn lot of stuff about us."

About the last of October some of the men went to Black Point and built a small shanty out of the wreck. First two men went and stopped down for three or four days at a time, then they would come up to Pot Harbor bringing along with them some young albatross. But before they would return back they would come to me and ask for a piece of pork and a little bread. I always gave it to them and told them that whenever they wanted anything they must bring up some young albatross with them and I do not think they ever came up without bringing something fresh.

About the first of November the penguins began [to] lay their eggs. When the men came up from Black Point one day, they brought with them about two dozen of eggs. About the middle of the week some of the men went down to Black Point where there is a large rookery of penguins. Before they started I told them if they got eggs enough we would have a big duff for Sunday dinner. So off they started ten men in high glee. When they came back each one brought one hundred and fifty apiece and they told the three men that was down there that Sunday I was going to make a big duff and they must come up and in the meantime I wanted them to gather all the eggs they could and cover them over with grass so the birds could [not] get at them.

The sheath bill, a white bird about the size of a pigeon, the only land bird on the island except the teal duck — if one should leave the eggs not covered, if there was a thousand eggs and he should leave them he would find a hole in each one. The sheath bill seems to do it for pure mischievousness and there is always large numbers of them around penguin rookeries. I have laid down on the ground about a dozen eggs and only turned around for half a minute and found them all with a hole in. Then again I have been off gunning and laid my gun on the ground while setting down and the sheath bills would collect around me by the dozens. They would keep drawing nearer and nearer; one, a little bolder than the rest, would walk up to the gun and cock his head on one side and try and look down the barrel. Then all the rest of them would crowd around the gun, examining the gun all over, looking in the barrel with first one eye then with the other eye. After looking the gun all over it ended by them having a fight for the possession of the gun.

Sunday morning: and with it came the three men from Black Point bright and early. One of the men said, "We heared that you was going to have a big duff so we thought we would come up and help you dispose of it, and we have run short of pork." Right after breakfast we went to work making the duff. In the first place I had the steward and cook put on the fire the half iron barrel full of water and all the old cans one could find to heat hot water in. Then I got a butter tub and broke about five dozen eggs in it and one of [the] men kept beating up as fast as I broke them. After beating up the eggs well, I kept putting in flour until the tub was half full of batter. Then I put it in a bag. As I was putting the batter in the bag one of the men that was standing around looking on said, "Captain, that will not be enough for all hands."

I said, "Tim, if that is not enough, next Sunday I would make a larger one but I think this will be enough and some to spare." When the water got to boiling I put the bag into the pot and we did not let [it] stop boiling for about five good hours: as fast as the water would boil away in the iron pot we kept filling it up with hot water from the cans so as to give the duff plenty of water. When the duff had been boiling about two hours one of the men by the name of Tim Reardon said, "Look, captain, it will burst the bag and if it keeps on swelling the pot will not hold the duff."

I said, "All right, Tim, let her swell and burst but keep a good fire under the kettle." Just before twelve o'clock the men wanted to know if the duff was almost done. I told them it would be done in about one hour. The men went to work and got a small piece of sail for [a] mess cloth and put it in the middle of the house. They then got all the tin and small pieces of boards for plates to eat out of and the cold salt pork that we had cooked the day before with a bucket of molasses; in fact, they made quite a picnic spread of it. At twelve o'clock I had [it] dished up. It turned out fine. The men had all they could eat and we had about a third of it left, the eggs making it so rich that the men could not eat as much as they thought they could. After the men got through eating I asked Tim Reardon, the man that thought I was not making the duff big enough for all hands, what he thought of it now. He said, "Captain, it was plenty big enough and I think it was the best duff that I ever ate." From that time as long as eggs lasted we had duff for Sunday dinner and the men would meet all together rain or shine.

At last the middle of November came around. The men and myself had been preparing for our expedition to Three Island Harbor, making moccasins out of elephant hide so as to save our boots. When the men came up from Black Point the next Sunday, I told them that I should go to Three Island Harbor on the twenty-first, Tuesday, if it was good weather and I thought that the men that was going had better stop at the house so if it was good weather we could make a good start in the morning. That night the men wanted to know what men I was going to take with me. I told them they must settle that among themselves as they all wanted to go [on] the picnic — all I wanted was five good men and the best pullers. They made it up among themselves that I should take Ed Carroll, John Reed, and three Portuguese — Frank, John, and Sebastian. Each one of us should take three pair of moccasins, one blanket, one pair of drawers, two pair stockings, [a] monkey jacket, and about one hundred fifty pounds of hard bread, six pieces of pork, and a small bag of peas to make coffee and [a] bag of salt. For a coffee pot we had an old paint pot that one of [the] men had found on the beach and for a frying pan we broke a piece of an old trypot about a foot square to fry our meat on and a boat axe. That completed the whole of our outfit, which we expected to last a month.

Tuesday the twenty-first came at last with the wind light from northward and thick fog. After breakfast the men wanted to know what I thought was best to do: to go or stop until the fog cleared off. But I told them we had better go as the wind was light and we could keep close in shore as long as the water kept smooth and if the weather became bad we could make a landing somewhere on the lee coast. They said, "Captain, you are boss of this job; we will do as you say."

At seven o'clock we had the oldest boat down to the water, all hands helping to get our things to rights except the three men that was down to Black Point. As I was about getting into the boat, the crew crowded around us wishing us good luck. I took each man by the hand and bid them goodbye. As I did so I asked them as a favor to me to keep from fighting and quarreling, they all giving me their promise that they would do so. When we had got into the boat, the men shoved us off the beach with a cheer, and, as we left the beach, they gave us three times three cheers and then they all run up on the high ground, watching and waving their hats until we was hidden in the fog. When we got abreast of Black Point the three men down there caught sight of [us] and they waved their hats and followed us down along the coast about two miles. When they found that they could not keep [up] with the boat, they waved their hats and turned back.

After getting around Mount Carmel Point[24] the fog became so thick that I could hardly see a boat length ahead and I could not see the breakers in time to keep clear of them. We had two narrow escapes from being capsized. When we got to a place called Sutter Bight[25] about halfway to Cape Digby from Pot Harbor, we had to land and make our first camp. Our camp was formed by turning [the] boat bottom side up with the bottom towards the wind and on the lee side of the boat we had some boards with us which we put under the lee gunwale so as to raise the lee side as high as [a] man's head and then [we] banked it up all around with sod earth. We had our door in the middle. We had a piece of canvas for the door. We had our chimney built of sods about two feet above the bottom of the boat. Then we filled under the boat with grass for our bed.

After getting our camp to rights we went looking for something fresh to eat. Three of the men went up on the high ground inland and they came back within half [an] hour with about a dozen young albatross and we went to work and dressed them for our dinner and supper. Usually [it] took six albatross for a meal, and all the time that we was on the lee coast where there was young albatross we ate that number at [a] meal. Ed Carroll we put in [as] cook. He used to take the albatross and cut off the breast and legs and then cut off all the fat and put the fat in our trypot frying pan and while it was trying out he would put in a little salt and when the fat got hot put in the meat. I do not know that I ever eat anything as nice as young albatross. Even when it is frying the smell is most delicious. But the best of all we had good appetites for sauce and good cold water to wash it down.

The next day the fog cleared off some but there was more sea running. Bright and early we was up and as soon as we could get something to eat we got away keeping close in shore just inside of the kelp, the wind keeping light until we got off Cape Digby. Off that cape we had to go off-shore about three miles to clear the breakers off the point. Before we could get inshore again, the wind [came] out strong from the northwest and the kelp was so thick the men had work pulling through it. The kelp was this string kelp; some of it was thirty to forty fathoms long. When the men would put their oars in the water deep, it would get around the oars and foul them up so that they would have to stop pulling and clear their oars. But after hard pulling, we got inside of the kelp but it kept breezing on stronger and stronger, the wind blowing off the land and a fearful sea running on shore from northeast.

I told the crew that we must make a landing in Royal Bay somehow or get blowed off-shore. Royal Bay is an open bay, all winds blowing into it from north around to west. Vessels cannot anchor in it unless the wind is northwest and then it is a poor harbor. We kept along to where they generally land. When we got abreast of the landing I had the men pull easy, just holding the boat from drifting, while I watched for a smooth time. While waiting I told the men what I wanted them to do. The man that pulled the bow oar, he must look out for the bread bag the first thing after the boat strikes the beach, and the rest of the [men] must look after the boat, for I told them that I expected to get full of water if I did not get capsized.

"Now, men, I want you to bend to them oars when I tell you to pull."

I kept edging in little by little until I got on the outer breaker and the crew held her there until a good smooth [one came]. When I gave the word to give away the men gave away with [a] will sending the boat spinning. I had all I could do to keep the boat straight on the seas. As the boat struck the beach a heavy sea struck her on the quarter and slewed the boat broadside on to the beach, filling her half full of water. But the men got the bread on shore all right. We got all our spare clothing and blankets wet. We did not stop for that but grabbed hold of the boat and rolled her over and got the water out and then hauled the boat up into the grass. Then we went to work and took off our clothing and wrung them out and put [them] on again.

The sun being out bright, we spread out all of our spare clothing and blankets on the grass to dry. After doing that we rolled the boat bottom up and banked up all around it with sods. Then some went to collecting grass, the grass being in large quantities and long. It took us but [a] short time to gather all we wanted. By the time they had the grass in, the rest of us had the chimney up and a good fire built. We had the fireplace built wide so all could get around the fire at [one] time. For fire wood we used the blubber from the sea elephant, which makes a very hot fire. The sea elephant was plentiful [to] hand so we did not spare them for fire wood. After starting the fire, before doing further work we took off our wet clothing, all but our underclothing, and dried them before the fire, turning ourselves around and around so as to dry the clothing we had on at the same time. After drying our clothing we began to think of something to eat. We all started off for our young albatross, all except Carroll, he stopping to get the cooking things [in] order. We did not go far; they was close by. By the time they was dressed and cooked it was dark and when we got through eating there was no albatross left to tell the tale, our appetites being so keen.

We stopped four days in Royal Bay, the wind being north all the time. In the meantime we used to pass the time by making trips, exploring along the shore and going inland. [Once] John Reed and myself [were going] along the shore when we came to a large water stream that was too deep to cross. I should judge it was all of one hundred and fifty yards wide and the current was running out very rapidly. Seeing such a volume of water running so swift, I thought we would follow the stream up and see where it came from. So we kept along the bank and followed the stream, and, when we had walked about a mile, the stream began to grow wider and deeper. When we had travelled about three miles inland, we came to a large lake about ten miles long and from three to five miles wide and full of small islands. The lake seemed to be very deep. I should think it would be a good place for trout, but I did not see no fish of no kind. In fact, we had no fishlines or hooks. After walking along the edge for some distance we returned to camp with some young albatross, for all around the lake there were large numbers, hundreds upon hundreds. Inland seemed to be the breeding ground. They had all left their nests.

The albatross builds its nest about two feet high in the shape of [a] sugarloaf with the top scooped out, and year after year they lay in the nest, and every year they keep adding to the nest. Some of the nests I have seen as high as three feet or more. The young albatross looks like a young gosling, only much larger in size when first hatched, and when six or seven months old their weight will be about thirty-five pounds. They are very fat about that age, and they have to get rid of that fat before they can fly. Along in October you generally find them off their nest, flapping their wings trying to fly. First off they will only fly a yard or two, then down they drop to the ground all in a heap. But they keep on trying until they accomplish the feat of flying.

While I had been gone the Portuguese went to Cape Digby Point, where there is a large colony of king penguins but they got no eggs, they not having laid, it being too early in the season. After eating our albatross and drinking pea coffee, we sat around the fire telling what we had seen and done throughout the day until first one then the other would fall off to sleep until there was no one but Carroll and myself left sitting up. After making a good fire and cutting some blubber in small pieces to put on the fire throughout the night, we turned in for the night and slept the sleep of the just.

The morning of the twenty-fifth did not look very bright, the wind being steady and strong from [the] northwest, but the sea had grown quite smooth through the night. We did not hurry, but took our time about getting breakfast. After we had eaten, I told the men we had better pack up our things and if the weather was the same when the sun got higher we would make a start. About nine A.M., the wind being about the same, I asked the men what they thought about making a start. Their answer was, "Captain, you know best. Whatever you think best we are willing to do."

"Well, men, I think we had better make a start and get as far as we can." We got the boat down, packed in our things and shoved off. I pulled to the head of the [bay] instead of going straight across, keeping inside of the kelp. By the time we got to the head of the bay it breezed up strong but I did not mind that as the wind was in our favor being on my quarter and the boat went flying down to the point on the southwest side of Royal Bay. When we got to the point, I pulled close in-shore as we could; that was no nearer than [a] mile and [a] half. As we was pulling along now and then a sea would come along that I thought was going to break outside of me. I would keep off and meet the sea head on and when the boat would get on top of those seas it would leave her so quickly that it seemed as if the whole bottom was stove in. [It] took most of my time to watch the seas. The wind became stronger and stronger in force until it became [a] strong gale.

I told the men that we must land. We had got down to the place we called the sand hills. I knew that the beach was sand and smooth but what troubled me was getting through the first two breakers. After getting through them two breakers, if the boat did not get capsized, we could wade to the beach, as the water was shoal. When a lull in the wind came I put the head of the boat in-shore. That brought the sea astern and the men gave away with a will. We went through the first two [breakers] all right, but the third one caught me on the quarter and slewed the boat broadside to the sea and filled us full of water and threw us all out of the boat. The boat went on towards shore and left us to paddle on-shore the best we could, but the boat got there first. As soon [as we] got to the boat the first thing we did was to get the bread bag out of the water and found the bread all right, the bag being No. O canvas, but everything else was soaking wet and ourselves looked more like drowned rats that had lost all of their friends.

After getting the water out of the boat we hauled her up on the beach out of the water. We then took our blankets and spare cloth-ing and wrung the water out of them and spread [them] out on the grass to dry, the sun being out bright at times. When we got our camp to rights we had [a] good fire soon a'going and when we got seated all around it we had a good laugh over our mishap. As one of the men expressed it, we looked like so many bull elephants coming out of the water. At sunset we got our blankets and cloth-ing but we found [them] still damp. We had to make them do for the night, [but] this was the only time on the trip we had to sleep in wet clothing.

When we had eaten our albatross we came near ending our trip. As Carroll was taking the frying pan from the fire, he let it fall and spilt the fat into the fire and by doing so caught the grass on fire that we had for our beds, and it made it lively for a few minutes until we got it put out; but the worst of it was we had to gather more grass for our beds. There was one of the Portuguese asleep at the time and he knew nothing about [it] until the next morning. The crew used to chafe him about dying when he went to sleep. He would take it all in good part; all he would say was, "Fire no come in my part; suppose he come in my place, I put him out damn quick." And there it would end with a laugh.

At the sand hills we stopped three days, and long days they seemed to us. On the twenty-eighth of November about ten A.M., the wind dropped off to calm and the sea as smooth as glass along the shore. We shoved off and continued our journey, keeping close in-shore, just keeping water enough to float the boat, and we made fine time getting along that night. Just [at] sunset we landed on Prince of Wales Foreland on the west side of Shoal Water Bay. We made our camp. We had just time to get the boat turned over and things to rights when it came on and blowed heavy with rain lasting three days.

There was one thing we missed, that was young albatross, there being none on the foreland. So we tried young penguin. The flavor of the penguin is very good fried in pork fat, but as we had so little pork we tried the fat of the penguin the same as we did the albatross but we found the fat not the same, the penguin fat having the taste more of fish oil while the albatross fat gives a nice flavor to meat that is cooked in it. But we ate the penguin all the same, but not with the same relish we did the albatross unless it was fried to crisp and it was better than no meat.

On December the second it was calm but thick fog; what wind there was was from N.E. We got away early, the tide being out at the time. We had to wade the boat off-shore a long distance and we got [a] good wetting before we started, as there was some sea rolling in on that side of the foreland. When we got around the foreland, the water was quite smooth; and it was a good thing it was smooth, for I had to keep close in-shore on account of kelp and the fog being so thick. The men was pulling with good will with a long swinging stroke; they began to be anxious to end the trip.

The foreland lies on the northeast side of Royal Sound; it is a ridge running north and south, about one thousand feet high, sloping off gradually east and north. In the west and south, it is perpendicular with small rocky beaches. It formerly used to be very green, but now since the rabbits have got there they have destroyed all vegetation. Three Island Harbor is in Royal Sound about twenty-five miles northwest from the south end of the foreland and is formed by three islands called Cat, Grave, and Hog islands, which gives it its name. There is good beating room between all the islands. But there is [a] large number of islands in the southwest side and at the head of the forelands, [where the] sound is about fifteen miles wide.

When we got around the foreland so that we could look up the sound to Three Island Harbor, I pointed it [out to] the men and said, "There, men, you can see where we have got to go to get to our journey's end."

The Portuguese said, "Captain, that no much too far; we get
there today suppose he keep good weather."

I said, "Yes, but we must not count our chickens before they
are hatched; but if it keeps calm I [am] in hopes to get there long
before night."

We got about half the way up the sound to [a] place called
Somerset Point, where the sound took a bend northeast, it still
being thick and calm. I asked the boat's crew what they thought it
best to do, whether to follow up on the east shore until we got
above Three Island Harbor or go direct from where we was. They
said, "How far is it around, captain, to pull on the east side?"

"I think about twenty-five miles and direct about ten miles."

"Then, captain, go direct."

"All right, but, men, we are running [a] great risk by going direct for if we get out in the sound in the middle and it should clear
off and come to blow strong, which it is likely to do at any time,
we would be in a very dangerous predicament as we have no land
on our lee and we cannot tell how the wind will come."

"Let us go straight if we can, captain, and we will get there,
you bet."

I said no more but headed the boat between Three Island
Harbor and the Rock of Dunder,[26] the men giving away strong.
When we got within three miles of [the] Rock of Dunder, the fog
began to lift and roll away to eastwards in [a] heavy bank. I could
see up toward the head of the sound the water was feather white. I
told the men that the wind was coming and that we must run off
and get among the islands on the west side of the sound. I kept
away a little so as to head into [a] place called Mutton Cove. We
could see the wind coming like a race horse, picking the water up
in sheets, carrying it high into the air, and hear the wind roaring
like [a] train of cars sending a short heavy sea before it. The men
did lay back on their oars and strain every nerve in their bodies,
making their oars bend like whipstocks, trying to outstrip the
wind; but it was of no use: we could not outpull the wind.

When the wind overtook us, we was within [a] mile of [a]
small island. And when it did strike us I thought it would pick
[us] up out of the water. The wind and sea filled us half full of wa-

ter. As luck would have it there happened to be [a] thick patch of kelp close by. I ran into it. We caught hold of it, making it fast all around the boat, the kelp making it quite smooth. We bailed the boat out and laid there about [a] half hour when we began to feel cold. While lying in the kelp, Carroll said it was a good thing we weren't out in the middle of the sound for if [we] had been we would have been blown out to sea.

"Yes we did not get half [the] wind when it struck that we would have got if we had been in the middle. I think there would be no one to tell what had become of us by this time. But there is one thing, boys, I shall not do hereafter this: I shall run no more risk." After a little the wind [began] to moderate. The men thought they could pull inshore [a] little farther so as to make smooth water of it; they could pull up to the Rock of Dunder. I said, "Yes, we might as well try [it] as to freeze where we are."

After getting clear of the patch of kelp, we went with the wind on the quarter until I got in smooth water. That brought the Rock of Dunder right into the wind's eye from us about five miles off. By the time we got into smooth water, the wind had lulled down to squalls, but the squall would come down strong at times. When we would see one coming, we would try and hold the boat until the heavy puff went by, but in some of the puffs we would go stern first and lose all we had gained. At last we got to a small island called Tumber Island, about one mile from Rock [of] Dunder. We got under the lee of it, for we could not land on account of it having [no] beach. But we done the next best thing: that was to let the boat lie close in out of the wind and make ourselves as comfortable as we could. We put our blankets around us and the sun shining bright we done very well. But us lying close under the rock, we got very little of the wind; only now and then a little puff would strike in.

I laid there about three hours. When we had a long lull, we left. By hard pulling [we] managed to get to Rock of Dunder, but found no beach to land on or haul up the boat. We made her fast good to the rocks and then we all took [a] stroll over the island. When we came back to the boat we had something to eat, hard bread and raw pork and it went down with a good relish. About

sunset the wind began to moderate down and [at] times it would
be almost calm. I said to the men, "Boys, it is about six or seven
miles more to go. I think we had better be going before it gets
dark."

"All right, captain, we are ready to try it again and see if we
can make a finish of it this time."

"All right, I am ready whenever you are, men."

When a good lull came, [we] struck out with good long swinging strokes. But it was hard work pulling against the wind and
short head sea, but we kept digging at it and we gradually gained
ground until we got under the lee of Grave Island when the wind
dropped off almost calm. We had no trouble getting to Hog Island. We arrived at Hog Island just before dark. We hauled the
boat up on the beach and made her fast good for the night. There
[was] an old house there belonging to Lawrence & Co. In the
house we found [an] old stove. We put it in the middle of the floor
and made a good fire. That night we had fried pork and hard
bread for supper. After eating we did not sit up long but made
good fire and then we rolled ourselves in our blankets, each one
picking out a soft board for a bed, we being that tired and fagged
out we could have slept most anywhere, even if it had been in a
water hole. I think no one did wake up until after daylight.

When I did wake the sun [was] up high in the heavens, and all
of the men sleeping sound and the fire all out. I got up, made the
fire, and then called the men. While Carroll was getting breakfast
ready I went to the back side of the island to see how the weather
was in the sound. I found it calm and I hurried back and found
breakfast all ready, some young penguin and lob hash[27] — we
made a fine breakfast. While eating I told the men it was calm and
we had better get across the sound as soon [as] we could; we
[would] not have [a] better time to [do] so. After breakfast I wrote
a letter and put it in [a] bottle and hung the bottle to one of the
rafters and Carroll put one in a bottle and hung it to [a] pole
driven in the ground. In the letter that [I] had written I stated what
had happened to the *Pilot's Bride* and where anyone could find me
[who] should come into Three Island Harbor looking for me.

Having accomplished [our] mission, we was ready to take the

back track. Seeing that all was right with the house, we left Three
Island Harbor for the east side of Royal Sound. When I got by
Hog and North islands, I had wind at times and calm but only for
short time. I kept well up among the islands at the head of the
sound. It was well that I did so for when I got about half-way
across it came on and blowed [a] strong gale at times. When the
wind would lull down we would pull from one island to the other.
Some of the islands was two or three miles apart and when I
would get half-way between them the strong winds would catch
me. Then we would have [a] hard time of getting to the next island. But after hard pulling and [a] hard day's work, we arrived at
[a] place across the sound called Malloy Point about four o'clock
in the afternoon, where we was going to leave the boat and walk
overland to Pot Harbor, a distance of about forty miles of rugged
and mountainous walking.

I had the boat hauled well back from the water into the grass
for I did not know when we should want to use her again or what
might happen. We turned her over for the last time to make camp
of her. As there was plenty of grass we filled under her full of grass
for our beds. As one of the Portuguese said, "He go sleep, him
good; I go walk much plenty tomorrow." That night we had penguin 'scouse, fried pork and fried young penguin for dinner and
supper, as we had plenty of pork left and about thirty pounds of
hard bread; and we used the last of our pea coffee for we could afford to have [a] feast.

The next morning we was [up] before the break of day and
had our breakfast cooked and we ate it by the light of the fire. The
morning was splendid, not [a] cloud to be seen in the heavens.
Getting through with our breakfast, we packed our things in as
small a compass as we could to be ready to start on our return trip.
After packing we let the boat down so that she should rest on the
ground all around. We put all the spare bread and pork under the
head of her, then banked her up all around with heavy sods of
earth to keep the wind from getting under her and blowing her
away. Then we covered her bottom all over with light sods to
keep the sun from drying her up. As soon as we had finished banking up the boat, each man took off his boots and put on mocca-

sins to walk in, made our blankets into packs, and strapped them
with our spare clothing on our shoulders the same as [a] knapsack
with our boots hanging one on each side. When we had everything to our liking, we started on our thirty or forty miles' journey.

As we gained [the] foothills we had [a] splendid sight. As we
gained one of the small hills at the base of the mountain the sun
was just coming out of its ocean bed, throwing its golden rays over
ocean, mountain, and valley. It was a grand sight. We could see
for thirty miles around except one way, that was northwest, the
mountain cutting off our view. There laid Shoal Water Bay with
its low land all around it, extending for miles with its hundreds [of]
lakes, the sun's rays striking the lakes and [the] old ocean making
them look like burnished gold. I could almost fancy I could
hear the lowing of cattle and hum of the human voice. I stood
spellbound for two or three minutes for the sight was something I
never expected to see at Kerguelen Island or ever to see again. The
charm was broken by Carroll saying, "Oh, what a splendid sight
— what would some people give to see this sight."

"Yes, Carroll, and travel half around the world to have one
look at it. Carroll, why am I not [an] artist? But come, this will
do, standing looking at the sunrise."

With one last look, off we started again, winding our way over
broken fragments of rock pumice. Then we would come to [a]
stretch of hard good walking and when we did we did not let the
grass grow under our feet. We kept working gradually up the side
of the mountain until we came to a deep ravine with a large water
stream running very swift of about fifty feet wide. It was a good
thing that we had no rain for the last few days, for if it had been
raining we could not have got across for the stream would have
been thirty feet deep. As it was we had to take off our clothing and
make them into a bundle and put them on top of our heads to keep
them dry. When we got ready we had to slide down the bank to
the edge of the water. When I got there the Portuguese said,
"Captain no go first; we go see how deep then captain come."

"All right, go ahead," and in they went like water dogs. The
first plunge they made they went to their armpits. I did not wait,

but followed them in. When I got in the middle I thought [it] would take me off my feet, the water was running so swift and it was icy cold. When we got out our bodies was like blueing bottles. We was not long getting on our clothing again.

After dressing, we kept on the tramp. Whenever we came to [a] steep hill, when we got to the top we would take a short rest. In one of our stops we had something to eat — bread and raw pork. Only those that have been in the same predicament can tell with what relish we did eat that raw pork. About every time we took one of them rests the men would ask me, "How much farther have we got to go, captain?" So they did this time while eating.

"Well, men, you know about as much as I do, but I think we have got about half-way by the looks of the land. You see that hill ahead? I think it is from us about five miles. When we get to the top of it, if I am not greatly mistaken we shall see Mount Carmel. When we do see Mount Carmel we have travelled two-thirds of the distance."

When we got to the top of the hill that I had pointed out to the men sure enough there was Mount Carmel and [the] old ocean around by Pot Harbor but we could not see Pot Harbor on account of a high ridge running between us, shutting off our view. As soon [as] the Portuguese saw Mount Carmel they wanted to know which way Pot Harbor lay from where we was.

"When you get to the top of that high ridge you can look down on Pot Harbor and perhaps you will see the house."

"Well, captain, you give us your bundle, we go ahead to house."

"No, boys, you have enough to carry your own. If you want to go ahead you can do so and we will be down to the house some time in the night if not before," for my feet was getting sore and I could not walk very well.

"But, captain, you more better let us fellows have your things to carry then you come more quick."

"No, you have all you can carry."

After resting for some time I told them we had better be on the move, so off the Portuguese started up the mountainside, they skipping like young goats from rock to rock, calling back to Carroll, Reed, and myself, "We tell them down to the house you're coming, captain."

It wasn't but a few minutes before they was out of sight behind a distant spire in the mountain ahead of us. Carroll, Reed, and myself being some years older, plodded along more slowly, making constant stops. Our feet kept growing worse all the time and more painful on account of the rough and sharp stones that we had to travel over and our moccasins began to play out, the sharp rocks cutting right through like a knife. But after a little while we gained the mountain and looked down into Pot Harbor and we could just see in the distance a white speck which we knew was the house and wished [we] was there. We took [a] short rest, then down the mountainside we went, which was more painful and tiresome than climbing to us. I thought we would never get to the bottom. When we did get down, oh, how thankful I was, for it seemed as if I could not lift my feet or move my legs a hundred yards further but must drop to the ground. At the first stream of water that I came to I took off my moccasins and gave my feet a good bathing and put on dry stockings for the rest of the way. I knew the walking was good. After resting about [a] half hour the sun was shining bright and there was no wind, it being dead calm. So we did not hurry; in fact, we felt so comfortable lying on the grass in the sun that we did not care to move but we kept a watch of the sun all of the time and when it got almost down to the mountaintops I told Carroll that we must make our move or night would overtake us before we got to the house. We shouldered our duds and as we climbed up the bank we saw some men coming from the direction of the house. We keep going along until we came to a small water stream and then we sat down and waited until they came up to us.

They proved to be Mr. Chipman, George Manice,[28] and two Portuguese; and they was well pleased to see us. As Chipman and Manice grasped each one of my hands, there rose a big lump in my throat so that I could not speak for some time. "Well, Chipman, how is the men and things in general at the house?"

"Everybody is well and things is all right at the house, captain."

"I suppose the three Portuguese got down to the house all right, did they not?"

"Yes, they walked into the house without anyone seeing them. The first words they said was, 'You fellows better look for captain, he played out hims feet sore he can no walk too much fast,' so four of us started to look you up."

"When did you see us first?"

"We seen you when you was coming down the side of the mountain." After answering and asking questions for [a] little while, we started for the house. I think the men would have been willing to have taken me on their shoulders and carried me the rest of the way, they seemed so pleased to see me back again with them. We arrived at the house about sunset, foot sore and just about played out. They all seemed pleased to see that we had got back again safe and sound. When the Portuguese got to the house, the cook and steward went to work getting supper and when I got to the house the supper was all ready. The first thing the steward said was, after the first greeting, "Captain, you must be hungry."

"Yes, steward, I think I can eat something good if you have some thing good to eat."

"Captain, we have the best the island can afford."

"What is that, steward?"

"Why, fried eggs, fried albatross, and slapjacks, and for drink pea coffee."

"Steward, that will do; it is a feast for a king." I think that I did do justice to the steward's cooking. While eating there was a running fire of questions asked [and] answered, all talking at once. But when I told them about landing at the sand hills and we getting capsized and the boat leaving us to paddle our own canoe and when we got caught in the sound by the gale of wind, there was grave faces and deep silence. I told the men that it would be a long day before I should undertake that journey again and the boat's crew echoed my sentiments. I did not sit up spinning yarns long after eating supper but turned into [a] blanket boy for the night. I do not think I had more than got under the blankets before I was asleep. The next morning I felt worse than any old truck horse, both lame and sore and for two weeks I felt that overland walk.

The next Sunday after getting back from my trip all the crew met together at Pot Harbor around a big duff, they having made none all the time we had been gone the fourteen days. I ask them why they did not make one for Sundays? They allowed they had no one to make the duff. "All right, men, from this time out we will have one as long as eggs last, but I am sorry that you did not gather more rockhopper eggs, say about ten thousand, for they would help out our bread and flour for eggs; 'scouse is fine eating." The four men still stopped down to Black Point most of the time, except whenever they wanted something; then two of them would come up and stop over night and go back the next day if the weather was good. There is one thing I must say for them; they never took nothing without asking for it.

Three weeks after getting back from Royal Sound, Charles [*sic*, George] Manice, one of the boatsteerers, and three Portuguese came to me and wanted [to] know if I had any objection [to] them going down to Black Point and building [a] house and living down there. They said, "Whenever you want us for anything we will come up and help you."

"Yes, you can go if you want to, but why do you wish to stop at Black Point?"

"You see, captain, we can live there without quarreling, which we cannot here. You know Mr. C. is always arguing about something and he thinks he knows everything and he tells such (d——d) lies."

"Manice, you have no need of getting in [an] argument with him. Do the same as I do, let him talk. You knew him of old; you came from the same place he did."

"Well, it is all right for you to think so, but not for us that have to bear the brunt of his talk."

"Well, let him talk; it will hurt no one."

"Yes, it will, for he is telling the men you are going to take them sealskins and sell [them] at the cape and put the money in your pocket and when the vessel comes after us you are going to let the captain know where the seals are and make the best terms for yourself and let the rest of [the] crew look out for themselves."

"I suppose you all believe what he says?"

"No, not all, but some of them do."

"Well, I shall take no notice of what he says for he can not help having a bad mind. Manice and you, Frank, John, Sebastian, you know what I told you when we got first on shore, that I would do the best I could for them if they would behave and them that did not I should let go their own way when the vessel came. As for looking out for myself, I should be [a] fool if I did not, for we would not have found the seal if it had not been for me, and Mr. C. should be the last man to find fault with me. You know Mr. C. and myself had words about looking for seal, he wanting to stop looking for seal and take oil."

"Captain Fuller, you need not be afraid. Most of the men will not take much stock in what he says for they do not like him well enough for that, but you must not tell him too much of what [you] want or [are] going to do, for he cannot keep nothing to himself."

Manice and the three Portuguese went down to Black Point the next day, where they built themselves a small house out of the wreck. They stopped there most of the time, only coming up whenever they wanted pork or molasses. The rest of the men at Pot Harbor, usually two and three at [a] time, would go down and stop [a] day or two with them. I think every one in Pot Harbor made them a visit but the mate, Mr. C.

About a month from the time Manice went to Black Point, one morning we was sitting in the house [when] we had [a] great surprise: who should come in but Charles Manice, bag and baggage. I said, "Manice, what is the matter? Got homesick? I thought you like living down there."

"Oh, I like it well enough but I [would] rather be with the crowd. It is too lonesome for me. Them three Portuguese do nothing but jabber Portuguese from morning till night."

"Well, Manice, why did not you learn to speak Portuguese? Then you could jabber with them."

"I cannot get that damn twist on my tongue, captain."

About [a] week after Manice came back from Hotel de Portuguese, one fine morning [I] thought I would like to take [a] trip down and see what they were doing. After breakfast Carroll and myself went. We got down [to] Hotel de Portuguese about noon. I found the men had built two houses about [a] mile apart and they were quite comfortable. They was much pleased to think that I had taken the trouble of coming to see how they was getting along. We put up at Hotel de Portuguese. As soon as we got there one Portuguese named John [said], "Captain, you want something to eat?"

"Yes, John, I should like to eat something if you have it."

"Ho, captain, you wait one minute. We get you something," and they did. They gave us fried young gull and slapjacks made from hard bread pounded up fine, sweetened with some of Masser Lawrence & Co.'s best molasses. They got their bread from a cask that had come from the wreck, also [a] half-cask of flour, but the flour had been opened and the flour was wet with salt water and it had fermented. With molasses it made first-rate slapjacks; at any rate, I thought so at supper time.

The next day I went to the American House, Odell, Reed, Reardon, and Fink Co. It was made of boards on the sides. For the roof they had the tryworks cover covered with canvas, but [it was] much smaller than Hotel de Portuguese. From there I went down towards Mount Carmel Point to see some of the wreck. When I came back I stopped at the American House and took dinner with Odell, Reed & Co. After supper they wanted me to stop all night with them. I told them no but would go and stop at Hotel de Portuguese for if it was a good day I should go back to Pot Harbor the next day. When we got back to the Hotel de Portuguese they had [a] lot of slapjacks cooked and fried young gull, so we took another supper with the Portuguese. The next day when I was about starting all three Portuguese said to me, "Suppose captain come down again, captain come stop with us; we very glad if captain would."

"Very good, men, when I do come down I shall not forget your kind invitation," but that was the last time that I ever walked down to Black Point.

One day I happened to be in the house alone with steward and cook and two men, James Magill and Thomas Flaherty. Magill wanted Flaherty to play him a game [of] checkers, so they sat down and played quite a number of games when they got to quarreling. Magill said, "Look here, Flaherty, if you can beat me at checkers I can whip you, [you] damn Irishman."

"Magill, you cannot whip one side of me."

"Flaherty, if you do not be damn careful what you say you will find that I will whip both sides of you."

"Magill, you had better try it on; you will find that I will give you as good as [you] send."

I thought it about time that I had something to say. I told them to stop quarreling and go outside and fight it out if they wanted to fight but you cannot fight in [the] house or quarrel, so outside you go. By that [time], Flaherty went outside and was walking away when Magill came out and called him a damn sneaking Irish coward.

"You come down on the beach and I will let you know if I am a damn Irish coward or not."

I thought nothing more about them, when the steward poked his head into the door saying, "Captain, they are fighting on the beach."

"Well, let them fight, steward, they will be better friends when they get through fighting." When I got outside they had got through fighting. Both had enough. Each had two black eyes and bloody noses. They was calling each other hard names. After they had got through, I told them they ought to be ashamed of themselves, to quarrel and fight over a game of checkers. But that was their last fight. They was good friends from that time out; they had more respect for each other after that fight.

The first of January: Reardon and Fink came up to Pot Harbor and when they was going they wanted [a] piece of pork and I gave it to them. After they had gone Mr. C. said, "Captain, if I was you I would not give them no pork. If they want pork to eat let them come to Pot Harbor with the rest of us."

"Mr. C., I rather they would be down to Black Point than up here, for you know they are a bad lot."

"Yes, I know they are [a] hard lot. That is one reason that you ought to keep your eyes on them."

"Well, they can do no harm down there and we will have enough of them when we get to Norton's Harbor."

The next time the men came up from the point it was Jack Reed and Odell. I happened to [be] up on the lookout place called Boar's Head. They came up there and wanted to know when I was going to move from Pot Harbor to Norton's Harbor. I said, "I am going the first good day in March."

"Captain, will you let us four stop at Pot Harbor or Black Point?"

"What four men do you mean?"

"Why, Reed, Rearden, Fink, and myself, Odell."

"No, you cannot by my consent for if anything should happen to you it would fall on my shoulders. No, I want all to go to Norton's Harbor."

"Be you going to take the sealskins with you, captain?"

"No, we shall have enough to carry without taking sealskins and the skins is all right where they are."

"Captain Fuller, I cannot see why we cannot stop and look after the skins. If a vessel comes from home she will have to come to Pot Harbor after them."

"Odell, you and the rest of your gang seems to think that them sealskins belong to them. Now I want you to understand one thing, that them skins I will take charge of from this [day] out and I do not want no more said about it and as for you men stopping you shall not if [I] can help myself, which I think I can."

They left me, feeling very sore against me, and when I got to the house they had left for Black Point. The last Sunday in February when Tim Reardon and Fink came up, I told them that they all must be at Pot Harbor the last day of February, "And you tell Reed and Odell that I want to leave the first good day for Norton's Harbor and they must be here on that day with all their things."

We had begun to prepare for our removal from Pot Harbor sometime before the first of March, and when the last day of February came it was raining and it rained all day. When night came the men from Black Point had not put in their appearance and some of the men had just said, "Captain, I think the men will not be up from Black Point," when in they popped, bag and baggage. That night about twelve o'clock it cleared off fine with light airs from eastward and in the morning it was beautiful. I had the steward and cook up before daylight and we had our breakfast long before sunrise and all our things packed away in the two boats. Just as the sun was coming up we shoved off for our journey from Pot Harbor to Norton's Harbor in Big Whale Bay, [a] distance of about forty miles. There was eleven men in each boat and with all of our things there was not much spare room. When we got outside of Boothead,[29] it was dead calm; the men gave away strong, each one taking a spell at the oars, whenever one got tired or cold. We kept close in-shore all the way up. The weather kept fine all day and we arrived at Norton's Harbor about eleven o'clock.

After getting something to eat we went to work putting up two houses, one for the officers and one for the men. While we was to work on the two houses, three of the men came to me one night after dark and wanted to know why it was that I wanted two houses. "Well, men, you know when down to Pot Harbor you men and Mr. C. was always in some foolish argument and I take this way to put a stop to all arguments."

"Captain, we should like to have you and the officers stop with us in our house, but if you think different, all right; for we know what Mr. C. is by this time."

"Yes, men, but you must remember there is more than you three in that house and some of them do not know when they are well off."

"How about us living? Have we got to do our own cooking, captain?"

"No, we will live the same as we have been living, but as soon as anyone finds fault with the cooking, I shall knock the steward and cook off and let them or him do their own cooking."

It took us three days to build our houses. We had plenty [of] boards and spars. We sodded the sides up to the height of five feet and then we covered the top with canvas, letting the canvas come well over the bank of sods and when it was finished the house was quite comfortable. After getting the houses all right, we went to work building a cookhouse as we had no place to cook in all the time we had been building our living houses. We had done our cooking outside, but we knew that we could not keep on doing so for this fine weather would not last more than a day or two longer. So on the fourth day after we came to Norton's Harbor we went to work, some of the men cutting sods of earth for the sides and some putting [up] the chimney, Jim Glass, the cooper, superintending the job. We made the door large enough to roll a cask of flour and cask of bread into the house. After banking up the sides and the ends, we covered the top with some canvas, making quite a comfortable place for the cooks.

It had two bad faults. The trouble was it would smoke when the wind was to the westward and the other fault was that it would collect soot and when cooks would be cooking rabbit stews, the soot fell down into it in big patches and it seemed to fall about the time the cover would be off. I used to tell the cook that it went for pepper and he would say, "Hot mon, damn bad pepper that." The steward and cook done the cooking. They would take turns about of two weeks at a time. When the cooks came out of the cookhouse after their day's work was done, they would look like Negro minstrels.

While we was putting up the cookhouse, Mr. Fuller and [a] boat's crew went after blubber to burn. They was gone two days. When they came back with about a boatload, we took the blubber and put it into a cask, first setting the cask two-thirds in the ground. We never used the blubber, only for cooking. The way we used the blubber was this. First we would start the fire with a little wood split quite fine. After the fire got going good, we would [take] a piece of blubber, mince it quite fine and lay it on the wood so as the flames could try it out, the oil falling into a hole under the grate upon the coals of fire. In that way we used very little wood, the blubber being more plentiful and giving more heat. We had one small coal stove which I had left there on a former voyage. I let the men have it on account of their house being [colder] than the officers'. We built a fireplace in the officers' house with an iron grate to burn hard coal in as we had about four tons on shore. It would burn fine in the grate, but we would only burn the coal when it was very cold or rainy weather. Other times we burned dry peat, the peat burning fine, it being full of vegetable matter and giving out good heat.

From the time we came to Norton's Harbor our food consisted of: breakfast, bread 'scouse and salt beef or pork. For dinner, rabbit stew. For supper, slapjacks, fried rabbit, or cold boiled pork. If any one wanted fried rabbit they had to fry the rabbit themselves as that was something extra, the cooks having nothing to do with it. Soon after getting settled down at Norton's Harbor, some of the men began to find fault about the way the food was cooked. I told them if they was not satisfied with the cooking they could do their own; I was satisfied. They tried it for [a] day or two, then they went to the cooks and asked them to do their cooking again for them. From that time out, we had no more trouble about cooking.

About the last of March one day I was in back from the house. Jack Reed came where I was washing out some clothing and we got to speaking about things in general when he said, "Captain Fuller, are you going to give any of the can goods that is in the cask and sugar to the men? The men say they understood you to say that we should fare alike."

"So we are, Jack, as near as I can do so, but who said that the men weren't going to have their share of the things?"

"Mr. C. has been telling Reardon that they was only put in for the officers and the men weren't going to have a damn smell of them."

"Reed, why is it that you men cannot take my word instead of listening to what Chipman tells you?"

"I do not know how it is, captain, but he makes things look so plausible to the crew."

"Well, Jack, I will tell you again what I told the crew right after the loss of the *Pilot's Bride*, that we should share alike in all we have got and you can tell the men so."

Soon as I got my washing done I went to the house. Mr. C. happened to be in. I said, "Mr. Chipman, why do you tell the men that they are not going to have any of the sugar or can goods?"

"Who told you that I said they weren't going to have none of the things?"

"If you want to know, it was Jack Reed that told me."

"Yes, that d . . . . negro is always telling something."

"Chip, you [are] doing your very best to make trouble between the men and myself. Why cannot you keep your jaw to
yourself?"

"Captain Fuller, they ought not to have a thing, and if I had
anything to say they should not [have a] (G——d D——n) smell
of the things."

"Yes, and thereby you would be in hot water with the men all
the time."

"I think I am able to take care of myself and fight my own
battles."

"Well, Chip, it did not look like it when Odell and Manice got
after you, all because you could not keep your tongue to yourself.
Well, from this time be [a] little more careful what you say and to
whom you say it to."

The next day I divided the sugar in two equal parts and gave
the men their part, telling them that they could do what they liked
with it. The green corn and peas and all can goods I keep myself,
giving it out to them once in two weeks as long as they lasted, the
crew being well satisfied.

About this time Odell thought that the officers was not good
enough for him to live with, so I told him he had better pack up
his things and go with his other chums, Jack Reed and Reardon. If
he did not I would kick him out. If any of the officers wanted anything done he must ask Jack Reed first. I stood it as long as I could
then I told him to go. We had a good deal of quarreling and some
fighting among the men.

About this time the Portuguese went to work making wooden
shoes, using wood for soles and canvas for the upper leather.
They made their own nails. They made me a pair. They was very
comfortable to the feet. The first time I put them on I went on the
rocks and the rocks happened to be wet. As soon as my feet struck
the rock they flew from under me as if I had been walking on ice.
Down I came on the broad of my back making me see stars, and
after that I did not venture on the rocks with them on.

One afternoon right after dinner Magill and Carroll went for a
walk. When night came [they] had not returned. We got our sup-

per and still they did not come long after dark. I began to get worried about them. Three of us started to look for them. We had got but [a] little ways from the shanties when [they] hove in sight, coming over a small hill. So we stopped until they came to us. As they came along we could see that they was heavily loaded. When they got down to where we was, I asked them what had kept them so late.

"Rabbits, captain, rabbits! They were so plentiful that we did not mind the time until night overtook us. We got fifty-two a piece and they are all large ones. They made a larger load than we thought when we first started with them and we had to stop to rest every little while."

"We was just coming to look for you. I did not know but something might have happened to you."

"Nothing but rabbits. In some of the holes we got five and six in each hole and it was fine fun hauling them out."

"Well, men, you must not be out walking around after dark on account of the many bog holes and if you should walk into one we should not know where to look for you."

About this time — the last of April — the duck began to moult. So one fine day when it was calm, I got [a] boat's crew and three spare men. We made it up that we would go to [a] place called Betsy Bay about ten miles from Norton's Harbor. We got away quite early and it did not take long getting to Betsy Bay. After getting there we could not find the large moulting hole, but we found two small holes. We got about [a] hundred and fifty ducks. Then we went into another bay called Cold Springs, a little to the westward, where we found the ducks plentiful in a small water stream. Where the stream emptied into the sea there was a small hole of water. In this hole was hundreds of ducks that could not fly on account of they had shed their feathers and the new ones had not grown out. Along the banks of the stream and around the hole was plenty of cabbage and Kerguelen tea or wild rose. The ducks live on the seed of the wild cabbage and they hide away in the tea from the bird called seahen.[30]

The seahen is a bird of dark brown or dun color looking like our crows and of about the same size but very fierce, they being a

bird of prey. The duck are [a] kind of teal. Their color is dark
brown and they weigh about one pound when fat. To distinguish
between the cock and hen, the cock bird is somewhat the larger
and on their wings there is strips of bottle green and topping the
green is a stripe of cream color and they carry themselves more
erect than the hen.

After landing and making the boat fast, we all got around the
hole to keep the ducks from getting overboard. As soon as we
began to catch them, they began to go up the stream and we after
them, some running along the banks and others wading in the
water up to their knees. Some times making a grab at some duck,
down they would go into some hole up to their waist. The ducks,
when the men on the banks would get ahead of them, would turn
and make a dive under the water and go down stream again. As
they would go by the men in the water, they would make a grab at
them getting two hands full of them. Then the men down stream
would turn them up stream again. We kept doing so until we had
cleaned the stream of all the ducks. When we got through we was
wet through with water and mud from head to foot. After getting
the duck together and counted we found we had about seven
hundred all told. We then put on what dry clothing we had and
by that time the wind had got around southeast with light rain.
We had come prepared to stop overnight but we being wet and [it]
likely to be bad weather for some days we thought it best to return
home to Norton's Harbor that same day and to get there before
night as we had [a] fair wind up. We killed three large elephant
that was on the beach and put it into the boat for ballast. Then we
reefed the sail and started back.

When we got out into the bay, the wind became stronger and
more abeam with a short chop sea, which made the boat throw the
water from end to end and when the water would strike us it
would go through like a knife. I do not know that I ever felt the
cold so much before, and it seemed as if we [would] never get to
Norton's Harbor. To cap it all, it came in thick and rain. After
sailing about two hours, we arrived home; but we was so used up
with the cold that we run the boat on the beach and the men from
the house had to unload her and take in the sail. Then all hands

got hold and hauled her up on the beach. But I was in a bad fix for all of [my] clothing was wet and I had no dry ones to put on. So Jim Glass, the cooper, let me have some of his and they fitted me like meal bags, he being a man of about six feet two and weighing two hundred and fifty-odd pounds. His reefing jacket came down to my knees, so one can tell how they must have fitted me, but they was better than none at all.

While we had been gone Mr. Fuller took the other boat and went up to a place called Mary Hester Brown[31] up in Big Whale Bay. He got about three hundred ducks and got back to the house long before bad weather came on, so that day's work gave us about a thousand ducks. The next day we took the innards out of the ducks and hung them up and for three weeks we eat nothing but ducks. We got sick and tired of them and was glad when they was gone.

About the last of May the men began to get worse to manage. There was fighting and quarreling about all the time and I had my hands full to keep order among the men. One morning, the first of June, it happened to be Jim Glass's turn to get up and make a fire and sweep up. After getting the fire going he looked outside and found one of the boats gone from the beach. He gave me a call. We all got up and dressed ourselves as quick as we could and went outside and [saw] that someone had taken our best boat. I had all the men called and found six men missing: Jack Melrose, Charles Fink, Tim Reardon, and three of the Portuguese; all three Americans [were] all three jailbirds.

I asked the men if they knew where they had gone and if they knew they was going. They said they did not know no more about it than [I did]. "Well, men, where do you think they have gone to?"

"Captain," one of the men said, "I think that they have gone to Pot Harbor, for Melrose has been talking about Pot Harbor the last two days. They have taken our best boat and left the old one that can hardly keep afloat. Well, captain, let us go after them and take the boat from them and let them stop where we find them and let them get along the best they can."

"No, men, we will not go after them for if they have gone to

Pot Harbor they will have to come back after something to eat. Then they will be in our power." After two or three days we did not miss them, for we had peace in both shanties.

About the last of July one moonlight night we was sitting in our shanty looking, talking of home when Carroll and Odell poked their heads in the door and said, "Captain, them fellows have come back for we see three men skulking over the hill."

"Well, get the men out and we will see if we can find the boat. Let Jim Glass, Mr. Fuller, steward, cook and Flaherty look out for the shanties. I think they have come for provisions and when we are asleep they intend to take what they want and then steal away again, and we must block their little game if we can. Carroll, where did you see the men?"

"They was coming up from the point and when they seen us walking they ran down behind the hill."

"Carroll, you and three men go over to King Harbor[32] and if you see the boat or men sing out. Odell, Reed, you come with Chipman and me; we will go down to the point."

Off we started on the run. When we got to the point, we found the boat lying off afloat with [the] painter made fast to a rock. While we was unfasting the painter we was joined by Carroll and one Portuguese. When we got in the boat, she smelt strong of sealskin. Reed said, "Captain, them damn rascals have been to Pot Harbor and have brought up them sealskins and have hidden them away somewhere."

"Yes, Reed, and I think I know where they have hidden them. Well, boys, we got the boat a good deal easier than I expected. I thought we would have to fight for her — let us get her up to the house as quick as we can." It being calm it did not take long pulling up to the house and hauling her up on the beach, and we took the sail and oars out of her. After doing so I called all the men together and asked them what it was best to do with the men. They had a talk among themselves. Then Carroll said, "Captain, you do what you think is right and we will stand by you."

"Men, I think we had better put them on one of the islands by themselves to keep them out of mischief and I think Atlas Island a good place."[33]

"Yes, captain, that is a good idea."

"We will give them their portion of the provisions and they must look after themselves from this time out."

"But, captain, [what] will they do for a house?"

"Well, they must do the best they can. We will give them some canvas and some cask staves. They can build themselves a house if they have a mind to do so or they can sleep outside. It is all the same to me."

"What shall we give them to cook in, captain?"

"We will give them a piece of trypot and some tin cans. Now, men, the officers will keep a watch tonight and if the men come to the shanty let them come in but do not let them know what we intend to [do] with them."

We all went into the shanties. After talking the matter over, we all retired except Chipman, he taking the first watch. In about an hour Chipman said, "Captain, the runaways are coming to the house."

I got up to have a look at them. Over our door there was a hole the width of the door that we had left open to let in fresh air and it made a very good outlook. [The] first one [who] came was Jack Melrose. He came sneaking around the corner of the cook house. We could see that he had something [in] his hand, which looked like a knife. He stopped and looked all around, then seeing the coast clear he beckoned to the rest. When they had came up to where he was they held a whispered conversation for a moment or two and then we could see that they had made up their minds to do something. We could see Melrose moving his hands and pointing towards the men's house. Once he made a gesture towards the officers' house. All at once they made a move, Melrose taking [the] lead, with Reardon and Fink following and the Portuguese bringing up the rear.

They kept their eyes on the door of the officers' house as they passed by. There being no light in the house they thought we was all asleep. As they went by we could see that Jack Melrose had a boat axe in his hand, ready to strike the first one that obstructed their passage to the house. Reardon and Fink had long skinning knives. When they got to the door, they stopped sometime before

they went in, Melrose wanting Reardon and Fink to go in first.
But they would not go in. At last Frank, one of the Portuguese,
got up courage to open the door and go in, the rest following him.
As soon [as] they went in we opened our door. We heard, "Where
did you in the devil's name come from? Where have you been this
last month?"

I heard Melrose say to Odell, "It is none of your (damn) business where we have been."

"You will find out that it is some of my business if you have
troubled them sealskins."

"Well, I can tell you that them sealskins are not at Pot Harbor,
Odell. They belong to me as much as they do to you, so you need
say no more." After that we could hear them laughing and talking
about their adventures long into the night.

The next morning after breakfast I had all hands called and we
got the six men outside and we held a court over them and Jack
Melrose was their spokesman. "Melrose, where have you been
this last month?"

"Captain, I do not know as it is any of your business where we
have been."

"What right had you to take that boat without asking and to
entice the Portuguese away with you?"

"We had as much right to the boat as you or any one else.
As for enticing the Portuguese away, they did not want much
enticing."

"Well, Melrose, it do not make a particular difference to me
where you have been; you had no right to put our lives in
jeopardy. If you had lost her we should have been hard pushed.
Now I am going to put you on Atlas Island, where you cannot
take a boat without asking for it. You can go peaceable, all right, if
not we shall use force but go you six men must. I will give you
one-third of the provisions. You can take care of it or not, just as
you like. When it is gone you cannot have no more. So get your
things ready and by that time we will have your provisions
ready."

After getting their things, they all six came down to where the
boat was. I was standing alongside of the boat when Melrose said,

"Captain, you have got the upper hand of us but if we had the crowd that you have got we would let you see if you would put us on Atlas Island or not."

"Melrose, you can be as saucy as you have a mind to, but you must keep out of the reach of this club or your head will come in contact with it." He kept out of my reach from that time out. The rest of the men coming down to the boat with the provisions put a stop to Melrose's impudence. We gave them some old canvas, two barrels of pork, about five hundred pounds of ship's bread, one hundred pounds of flour, an old pot, a piece of trypot for [a] frying pan, and some wood, and I gave them some matches. When we got the things into the boat I made the runaways three of them get into each boat. They made no resistance, but [got] in and took their places without a word. We put them on the island and after landing their stuff I said to them, "Men, I will come and see you about once a month."

Melrose said, "You go to (h—— G—— D——). You fellows think yourselves damn smart, but you will see who is the smartest one of these days."

"Well, Melrose, we are [a] little too smart for you this time," Reed said, "And you think you have hidden them sealskins where we cannot find them, but you are mistaken."

"You go to (h——l) you Cape Town Negro," was all he said.

After landing them I went back to the shanty, and Mr. Fuller took the other boat and down the bay to the cave that was on Norton's Island and when he came back he brought one hundred sealskins that the runaways had hidden away in the cave. We took a small cask and salted them down in it and headed the [cask] up knowing that the skins — at least hoping so — would give us no more trouble. But in that we was mistaken.

For about a month after landing the runaways on Atlas Island, we lived in peace, when one calm day we happened to be on the hill back of the shanty looking down to Atlas Island when we seen a small boat shoot out from one of the small islands and Chipman said, "My God, captain, Melrose has made a small boat."

"Yes, I see he has and it is nothing more than I expected." We watched until they pulled into [a] bight in the island out of sight.

That night when we was setting around the fire talking the conversation turned [to] Melrose and [the] boat. Manice said, "Captain, what do you think Melrose made his boat out of?"

"That is more than I can tell but I think he has made it out of that canvas that we gave him for his house. You know Melrose has ingenuity enough in him to build a boat of that canvas and them cask staves, him and that Portuguese Frank."

"Well, suppose he has used the canvas for a boat, what do they do for shelter?"

"Oh, probably they have found a cave on the island; I know there is one on the island. And I tell you what, gentlemen; I expect to have a visit from Melrose some calm moonlight night and if he can do us any damage he will if he can."

"Oh, no, captain, he will not dare to come from there in a canvas boat."

"Well, you will see in time."

One morning I was taking a stroll and I thought I was the only one away from the shanty. I was trying to get away from my own thoughts. When I got about three miles from the shanty I heard voices rising [in] high words. There happened to be a ridge between them and me so that I could not see who the men was. I stopped to hear what they was quarreling about. I made out one of the voices to be Odell's. I heard him say, "(G—— D——) you, what did you do with them sealskins?"

Then I got hold of the other voice. It was Tom Flaherty. I heard him say, "Damn it, man, I tell you I know nothing about the skins."

Then Odell said, "Tom, you are the G—— D—— liar and if you do not tell us where they are we will knock your brains out with our clubs."

Flaherty said, "You and Jack can knock out my brains but I tell you, man, I know nothing about the sealskins."

"You are the G—— D—— liar that ever God let live, Tom Flaherty," said Odell and Jack Reed, "and we will give you just two minutes to tell before we come down on your head with our clubs."

I thought it about time for me to interfere. I got to the top of the ridge as quickly as I could. There was Reed and Odell with Tom Flaherty between them with their clubs raised ready for striking. With one glance I took it all in and I gave one shout, and as soon as Flaherty heard the sound of my voice he broke away from them and made a beeline to me, the other two following sharp upon his heels.

"What is the matter, Tom?"

"By jobs, captain, Odell and Reed say that I have stole their sealskins and [the] two of them says that if I do not tell them where they are they will break me skull, but devil a bit do I know about them."

"Well, Tom, I think they will not break no one's head today; but I should like to know where they got sealskins from."

"By jabers, captain, they must of took them from the cask at Pot Harbor before we came to Norton's Harbor."

I said to Odell and Reed, "How did you come by sealskins?"

They said, "We took them out of the cask at Pot Harbor and we brought them up with us and we buried them and now we find them gone."

"Who else knew of the skins?"

"No one knew of the skins but us two and Tim Reardon and Fink."

"Well how do you know that Flaherty stole them? Did he also know where they was buried?"

"We think it was him that took them, for [we] went to where they was buried this morning and found them gone and Tom was down here all by himself."

"You might say the same about me, and if Flaherty did steal them he has [as] much right to them as you. How many skins was there, Flaherty?"

"They say nine skins, sir."

"Well, Odell, I don't think that Flaherty took them and [he does] not know where they are."

"And be jobs, captain, I wish I did know where they have gone to."

"Reed, did you ask any of the rest of the men about the skins?"

"No, we thought we would keep a watch and see if we could find out before saying anything."

"Yes, you would have killed Tom Flaherty for nine sealskins if I had not come along when I did, when he knows nothing about the skins."

"We did not intend to harm him, only to scare him."

"Yes, your actions looked like it, with your clubs raised ready to knock Tom in the head. Hereafter I shall not trust you two for I think you two are capable of doing most anything that is bad. But come, Flaherty, and you two can come, we will go to the shanty and see if any of the men knows anything about the skins."

As we went along Flaherty kept close to me; the other two laid behind some ways. Flaherty said, "Captain, it was a good thing, be jabes, that you came just in time. Them two sculpins would have beat out my brains for they had two clubs and I had nothing but my two fists and be jabes you just saved my life, you did."

"Why, yes, Flaherty, I think they meant what they said to you and they would have done it too if I had not got along when I did. They could have done the job quite easily and shoved you in some of them bog holes and no one would have been the wiser."

"You see, captain, I was only taking a little walk to have a look out of the bay to see if I could see a vessel from home, and I had just turned around when the two cursers came upon me [at] one time. I thought I was a gone Irishman and by jabers when you sang out how me heart did leap for joy."

"Well, Tom, you see this walking stick and you see that bit of iron on the end? I had the cooper put it on to dig out rabbits with and to defend myself with. If they had struck you one blow, one of them would have felt the point of my stick and that one would have been that Cape Town beachcomber Jack Reed for he is the one that makes the balls and gets the other fool to fire them off for him while he keeps out of harm's way. Now, Flaherty, let me give you a little advice. Never leave the shanty without a good walking stick and if them two fellows attack you again use your stick and strike out with a will."

When we got [to] the shanty I walked right into the men's

shanty. The other three followed me in, and I said, "Men, which of you have taken or stolen Odell and Jack Reed's sealskins?" They all looked at me with open mouths. They was speechless. At last Carroll said, "Odell and Reed's sealskins? I should like to know where they got their sealskins from."

"No matter where they got them from, what I want to know is, do any of you know anything about the skins?" Their answer was, "No, we know nothing about them."

"Now, men, you want to know how they came by [the] sealskins. Well, they stole them from the cask at Black Point. They brought them up when we came up from Pot Harbor unbeknown to us and buried them. Now someone has stoled them from them."

Carroll turned to Reed and Odell and said, "I think if anyone stole the skins it was you, and I am (d——n) glad to think someone also has stolen them from you."

"Yes, perhaps, Carroll, you know where they are or who has taken them."

"Well, if I have taken them you will not get them again, Odell, and I hope the one that has taken them will keep them."

"Well, it is of no use of you men quarreling about them skins for I think I know who has taken them."

"Who, captain," said they all at once, "who can the fellow be?"

"Why if you will only stop to think you will know as well as myself who has taken them."

"No, captain, we cannot think who has taken the skins."

"Well, I think Jack Melrose has paid us a visit and he has left his mark on someone and the one is Odell and Reed."

"If Melrose has been over and taken them skins I will be even with him one of these days," said Odell.

"Odell, Melrose does not care a button for you or Reed either. What makes me think Melrose and his gang [took] the sealskins is this. You know Reardon and Fink knew all about the skins and where they was stowed away and they told Jack Melrose and the rest so to get even with Odell and Jack Reed for helping to put them on Atlas Island. They have taken the skins."

"Surely, captain, Jack would not come over in that boat which we saw them in yesterday."

"Why, yes, you let Melrose alone [and] he made that boat on purpose to pay us a visit."

"Well, captain," said Reed, "Why do you think it was Melrose? The skins was all right yesterday."

"Well, you [see] last night it was calm all night and this morning I was up long before anyone and I went to the spring for a bucket of water and I saw some footprints between [the] high and low water marks. I thought nothing about it at the time but as things has turned out it must have been Melrose and some of his gang."

"Captain, how did you find about the sealskins being taken away?"

"You see I was taking my usual walk when the morning was fine when I came upon Odell and Reed threatening to break Flaherty's skull if he did not tell them where he had hidden the sealskins."

"And boys, they would have done it too if the captain had not come along when he did," says Flaherty.

"I think you two fellows had (damn) sight of gall to talk of breaking folks' heads for nine sealskins," said Carroll.

"Yes, Carroll, I think so too, and from this time out I shall go armed, ready to defend myself against them two men."

About a week [after] that rumpus, one fine morning [I] said to the men that was around that I wanted [a] boat's crew to go over to Atlas Island and see how Jack Melrose's gang was getting along. They all wanted to go. I told them no, all I wanted was [one] boat's crew for I did not want Melrose's gang to [think] that we was afraid of them. So I took Flaherty, Magill, Carroll, and two Portuguese. When I got over there I found they had built their shanty under a cliff of rocks. The cliff was in such position that [it] formed [a] half cave. It was quite dry and comfortable. They had only used but little of the canvas that I gave them and that little for [a] door.

When I arrived abreast of the shanty I sung out. They all came out with sullen visages and wanted to know what we wanted of them, but I told [them] I should like to know how they was getting along. Melrose was the spokesman. He said, "We are getting along all right, no thanks to you folks."

"Yes, I see that you have as much lip as ever and I think you paid us a visit [a] few nights ago and left your card."

"What do you mean by leaving my card?" with a grin playing over his face.

"First, Melrose, where is your boat that you have built and where do you keep it?"

"The boat is all right, captain, and where it is is best known to ourselves."

"All right, Melrose, I should like to know about them sealskins that you and your gang has stolen from Odell and Reed. Do you know that they was going to kill Flaherty because they thought he had stolen them?"

"You said, captain, that I stoled sealskins from Jack Reed and Buck Odell. I should like to know where they got the skins from. Them skins is [as] much mine as theirs, captain. You think this gang all damn rascals, but [you] will find Reed and Odell [a] (G—— D——) sight worse, only they dare not do it openly. Where do you think they put that cask of skins before we came up to Norton's Harbor, captain?"

"That is more than I can tell, Jack, but I always thought the skins was left in the cask where they was found."

"No, captain, they weren't. Them two good men of yours with the help of Reardon and Fink took the skins out of the cask and carried them inland about one mile and buried them."

"What did they do that for, Jack? They must know that they could not get them on board of a vessel without my knowing of it."

"They thought they could get you to give them so many for telling where they was put, so I got it out of Fink where the skins was hidden away. I thought, captain, the second thief had better right to the skins than the first thief, so we took them and brought them up and put them in the cave."

"Yes, Jack, them skins will cause no more trouble. They are in a cask at the shanty. Thanks to you all, I can keep my eye on them. Well, Melrose, I am glad to see that you and the rest are getting along all right and that you keep yourselves clean. I have brought you over a few bars of soap," and I turned to the men in the boat and said, "Men, I think we had better be getting back to the shanty."

"Hold on, captain," said Melrose, "All the rest of the crew are curry-favoring with you; so will I. I will make you [a] present of them nine sealskins."

"All right, Jack, I will accept them, only I do not want Reed and Odell to knock me in the head because I have got them as they was going to do to Flaherty."

"No danger of that, captain, unless they happen to get you alone."

After putting the skins in the boat, I said, "Jack, I thank you for the skins. They are mine now, but they will go into the cask with the rest of the sealskins. Let me give you [a] little advice. Do not come over to our shanty again for you do not know what may happen to you."

"When I do pay you a visit again, [you] blokes will know nothing of it and if them two good men of yours should come in my way someone may get hurt."

"Well, Jack, you can please yourself about coming but that is my advice to you."

When we was pulling back [one of the Portuguese] said that Melrose's gang had their boat just around the point, [a] little ways from the shanty. The men asked how they like living on the island. The Portuguese said they was sick of living over there by themselves and that Jack Melrose tried to boss them too much. After getting back to the shanty I had Glass, the Cooper, put them nine sealskins in the cask with the rest. When I took the skins out of the boat Odell and Reed had smiles all over their faces, but when I told them that Melrose had given them to me and that I was going to put them with the rest of the skins they pulled a very long face but said nothing.

One morning about [a] week after I had been over to Melrose, we [saw] a boat coming from Atlas Island with two men in it. It was quite early, before we had breakfast. It happened to [be a] fine morning, almost calm. When the boat got near enough to distinguish the men, we saw one was Reardon, the other Portuguese Frank. When we made them out all hands was down to the water's edge and we all was speculating what had happened; some had one thing, some another that had happened to some of the gang, but all was wide of the mark.

As soon as the boat got near enough Reardon said, "Captain, I should [like] to speak to [you] alone."

"Well, land and you can speak with me."

As soon as they had landed Reardon and Frank came straight to where I was standing, saying nothing or even looking at the men when they was spoken to. I said, "Well, Reardon, what is it you want of me?"

"Captain, we want you to take us two back for we are afraid to stop over there with Jack Melrose. He has threatened to kill us two."

"How is [it], Reardon, that you have fallen out with Melrose and he wants to kill you? What have you done to him?"

"Well, captain, you see Frank has too much to say and he wanted me to kill Portuguese Frank when he was sleeping. He gave me a sharp pointed knife to do it with."

"He wanted to make you the cat's-paw to pull the chestnuts out of the fire, did he not, Reardon?"

"Yes, sir, but I wasn't quite fool enough for that. He said, 'Tim, you can do it quite easily by giving Frank a dig in the side with the knife and no one will be the wiser.'"

"Reardon, it was a good thing you did not do what Jack Melrose wanted you to do when you know it is [a] crime to do it. If you had done what Melrose wanted you to do, you would have been hung and Melrose would have been the one to blow on you."

"I know that, captain, but Melrose could not get me to [take a] life."

"Reardon, what started Jack Melrose to quarreling with you all?"

"The two other Portuguese took sides with Frank and that only left Fink [and] myself on Jack's side, and the Portuguese would have nothing to do with Melrose. So he asked what side I was on. I told him I should go with the Portuguese. That made him mad and then he threatened to kill Frank and me."

"I do not see why you four men should be afraid of Melrose and Fink."

"Captain, you do not know Jack Melrose as well as we do. Captain, if we should be walking along the edge of a cliff he could give us a push off and then he would swear that we had tumbled off. I would not trust Melrose. So Frank and me made it up between us that we would take the boat the first calm spell and come over to you."

"Did the other two men know that you was coming over to me?"

"Yes, sir."

"Why did they not come with you?"

"The boat would not hold all of us, sir."

"Well, Reardon, you and Frank can stop with us as long as you behave yourselves and when you do not behave I will put you on one of the other islands."

"Captain, you need not be afraid of us not behaving. We have had enough to last the remainder of the time we are down to Desolation. And, captain, we thank you for taking us back again."

"All I ask of you and Frank is to behave and keep out [of] all scrapes."

I went back to the rest of the crew and told them what had happened and that I had agreed to take Reardon and Frank back again. They thought I did what was right. While I had been talking with the two men the crew had hauled their boat up on the beach and was examining her. The longest piece of wood in her was not over two feet long. They was fastened together with ropeyarns, not a nail in her. She was covered [with] canvas. After we had looked her all over I had her broken up.

The next day I had another talk with Reardon and they gave me to understand that [the] two Portuguese over to Atlas Island would have [a] hard time of it with Melrose. Reardon said, "You see, captain, Jack Melrose will be mad enough when he finds us gone with the boat to kill the boy and Manuel." I told him I should go over the first good day.

About [a] week after them two came back, one fine morning I told the men that I wanted [a] boat's crew to go over to Atlas Island to see [how] Jack Melrose and his gang was getting along. I had no trouble about getting a boat's crew; even Portuguese Frank wanted to go over, but I told him he had better stop where he was. When I got over there Melrose and Fink were walking up and down under the cliff. The two Portuguese was nowhere in sight; as the boat came alongside of the rocks I told one of the men to get out and hold on to the painter, neither Melrose or Fink saying a word or offering to come near the boat. As soon as the man got out of the [boat], I asked Melrose where the two Portuguese, Manuel and the boy, was.

"They are all right, Captain Fuller, no thanks to you."

"Melrose, I think you had better keep a civil tongue in your head and if the Portuguese are all right I want to see them, or have you put them out of the way the same as you wanted Reardon to put Portuguese Frank?"

"They have been telling you that damn lie, have they?" said Melrose.

"Yes, Melrose, and what is more I believe what they say for I think you are capable of doing most anything that is mean and low."

"Did you come over to tell me that, Captain Fuller?"

"Yes, Jack, I came over to tell you that and to see if the Portuguese was all right. I am thinking of taking them back with me, seeing that you cannot get along without quarreling with them."

"I don't care a damn what you do with [them]. I shall be glad to have you to take them off my hands."

"Yes, Jack, I suppose you do not care what becomes of them now you got all you can out of them and after [getting] them into [a] scrape." I saw that it was of no use to talk with Melrose, so I told one of the boat's crew to look into the shanty and tell them two Portuguese that I wanted them. In [a] few minutes they came down to the boat. I said, "Manuel, what is the reason you stop in the shanty?"

He said, "Jack Melrose after Tim and Frank take him boat Jack tie boy and me when boy and me sleep and when him see boat coming to take him off and he say [if] you go out speak [to the] captain and tell him I tie you up and kill you when captain go."

"So Melrose has had you in irons, has he, and you have had enough of Melrose, have you, Manuel?"

"Yes, captain, me like go with you."

"Well, do you think you two can behave yourselves if I take you back with me?"

"We do what captain want, me no go in boat unless captain speak go."

"Well, get your things and while you [are] about it get Reardon and Frank's things and put them in the boat." When they had got in the boat I turned to Melrose who was standing a little ways off and said, "Jack, I will leave Fink with you. I think he is about fool enough so that you will not want to kill him and if you should get to fighting and should kill one another it will not matter much. But there is one thing: I shall look to you for his safety."

"You had better take him too, captain, if you think so mean of me."

"Jack, how can I think otherwise after you wanting that fool of a Reardon to kill Frank when he was asleep, but you can thank your stars that Reardon was not so soft as to do what you want him to do."

"Captain Fuller, Tim Reardon is a damn liar. He told you that to get [on the] right side of you and when I get hold of him I will break his damn neck for him."

"No, Jack, I do not think you will hurt no one, much more [that is, much less] break somebody's neck. And, Jack, there is one thing more I want you to bear in mind. Don't you undertake to come over to our place again, for just as sure as you do and I catch you I will put you under the ground."

"You had better catch your rabbit before you eat him, Captain Fuller."

"I will catch him and eat him too, don't you forget it, Melrose. I shall keep watch calm nights after this. So if you want anything you can make a fire and I will come over as this will be the last time unless you want something, for I am in hopes there will be a vessel from home in about two months. So take good care of your provisions and so long."

As we was pulling I asked the two Portuguese how it was they had got out of Jack's good books. Manuel said, "Captain, Jack Melrose he no want do nothing. He no dig rabbit. He sit down in house all time eat, do nothing. Frank, me and boy no like that. Frank he speak Jack, 'You go dig rabbit, Jack.' He get mad like the devil. He speak Frank, 'Suppose you no do what I speak you, me kill all you damn Portuguese.' And then Frank speak Jack, 'You better look out yourself if try to kill me I try kill you first and I do it damn quick too.'"

"So Jack wanted you all to wait on him?"

"Yes, captain, he speak all time, 'You, Manuel, you, boy, go get bucket water and (G.D.) you fellows go dig rabbits.'"

"Did he make Tim Reardon and Fink do the same as he did you Portuguese?"

"Fink, no; Tim Reardon, yes, because Tim no do what Melrose want him to do."

"What was it that Jack Melrose wanted Tim Reardon to do and Tim would not do it for Melrose?"

"Captain, he want Tim to kill Frank when him sleep so Tim he speak me."

"Did you hear Melrose when he asked Tim?"

"No, captain, Tim he speak us fellows."

"Well, Manuel, you chaps did not have so good [a] time with Melrose as you did with us."

"No, captain, Jack him one damn rascal; me no like. Captain, when Frank and Tim tooked the boat, the next day when me sleep Jack and Fink him tie my hands and feet and then they catch boy and tie him too."

"What did he tie your hands and feet for? Was he afraid you would get off the island without a boat?"

"Me don't know, captain, what him tie us for. When he see boat coming him afraid he take him off damn quick."

When we got back to the shanty the men wanted to know what Jack Melrose and Fink had to say. I told them Fink had nothing to say. Melrose did all the talking. He had as much lip as ever and when I taxed him about wanting Reardon to give Frank a dig in the side with a knife he only laughed and said Tim was a damn liar.

While we was getting dinner, some of the boat crew was telling about the scene between Melrose and me; we could tell by the peals of laughter that came from their shanty. The event gave the crew something to think and talk about. But as for myself no one but the good God above knew how it worried me and how anxious I was getting for the time when we might expect a vessel from home.

At times I was so anxious that I could neither eat or sleep. How I did envy the officers when they turned in for the night. They seemed to drop off without any trouble. As for myself, sleep would not come. It was turn and twist from one side to the other until the small hours in the morning. Then up I would get. But I could not stop around the shanty, so off I would tramp until night, then back to the shanty I came. In the morning as soon as I could get something to eat, off I would go again for another day's tramping over the mountains, anywhere to get away from the men and my own thoughts, for I began to see that I should have trouble with Reed and Odell. They began to throw out hints [that] they weren't going to do this and they weren't going to [do] that to the officers, and I knew they would be throwing the hints at me sooner or later, and when they did there was going to be trouble the same as it was this morning.

I had some words [with] Tom Flaherty and he became offensive in talk. Chipman was standing alongside of me. He happened to have his walking stick in his hand and it was a good stout one at that. When Tom began to use offensive words, Chipman brought [it] down on his head and it stopped his argument right off. I thought it was going [to] bring on a fight as some of the men began to have something [to] say about the trouble as high words began to flow. Glass, the cooper, Manice, and Mr. Fuller came around ready to take a hand, only waiting [a] word from me. The men had sense enough to see that they would be roughly handled.

They soon cooled down when they found we had just as ready fight as do anything else. From that time out I had no fears. I knew the officers would back me up in anything that was right. The four runaways did not have much to say and they behaved very well, but I could not tell how long it would last.

About the middle of July one night as we was setting around the fire, I told the officers that someone must go after some blubber for the cooks told me that our fire wood was getting short.

"I will go," said Mr. Chipman.

"All right, Mr. Chipman, you had better go tomorrow if it is fine and you had better go as far as Port Palliser while you are about it and get some young albatross."

"All right, captain, I will get a boat's crew and tell them to get ready."

In [a] few minutes he came back. I said, "How is [it] going, Mr. Chipman?"

"I shall take Odell, Reed, and three Portuguese — two Johns and the boy."

"I think they will be all right. After you come back I shall send Mr. Fuller down to Pot Harbor to see what Melrose and his gang has taken of the stuff that we left there when we came up to Norton's Harbor." The next day happened to be [a] fine day, so right after breakfast Mr. Chipman and [the] boat's crew started for Morgan's Bay and Port Palliser. They expected to be gone about a week or ten days.

One day about a week from the time the boat had been gone — we was expecting them back every day — I happened to be down on the point looking out of the bay by myself. I was down there in hope to see a vessel from home when about three o'clock in the afternoon I [saw] the boat returning. I could see that they was pulling a good long swinging stroke and I expected they would come along by the point where I was, but instead of that when they got within about a quarter of a mile of the point, all at once the boat kept away for Atlas Island. I kept watch of her until they landed at Melrose's shanty. I could not think what was up, knowing that Melrose liked Chipman as well as the devil did holy water. To say that I was surprised would be no name for it. So off to the shanty I started, puzzling my head why Chipman went to Atlas Island. When I got down to the shanty, Manice and Carroll came running to me, saying, "Do you see the boat over to Jack Melrose's shanty, captain?"

"Yes, I see the boat there and I saw them when they went there; I was down on the point."

"What do you think they went there for, captain?" said Carroll.

"That is more than I can tell, Carroll, but I do not think Chipman would go there of his own free will."

"My God, captain, I bet my head for a football that it [is] some of Jack Reed's and Odell's works," said Manice.

"What makes you think so, Manice?"

"I can't tell what makes me think so, but there [is] something that tells me that we will find them two devils at the bottom of the boat being over to Jack Melrose's shanty."

"Well, my boys, we shall soon know; the boat is coming."

"You mark my words, captain, and see if I am not right," said Manice.

They was sometime pulling over to our shanty, for the wind was blowing strong from northwest and that made a sharp chop head sea to pull against. All hands was down on the beach waiting anxiously for the boat to come to hear the news. At last they arrived. We all took hold and unloaded the boat of her blubber; then we hauled the boat up alongside of the other boat. I said nothing to Mr. Chipman about being over to Atlas Island until [the] boat was out of the water. Then I said, "I was down on the point; I saw you when you went over to Atlas Island."

"Yes, I was forced to go by them two damn scoundrels Reed and Odell. You can bet your bottom dollar that I did not go of my own free will."

"How was it, Chipman, the Portuguese went with them?"

"They talked the Portuguese over somehow and after they got them to say they would go it was all done in that damn Portuguese lingo that I don't understand [a] damn word. They threatened to knock me in the head and throw me overboard if I did not go to Atlas Island."

"What did they want over to Atlas Island, Chipman?"

"Captain, by what I could make out they wanted Melrose and Fink to join them in taking the best boat and [going] to the Cloudy Islands after seal, whether you like it or not."

"What did Melrose say to them?"

"Melrose told them to go to hell. He hardly would speak to them at first. They tried coaxing but it was of no use; Melrose would have nothing to do with them. You see, captain, we had about thirty young albatross. They took them out of the boat and tried to coax him with them albatross. They tried their best to get Melrose to join them. At last Melrose told Odell that he would see him in Hell first for they was two bigger damn rascals than he was. They was about an hour persuading him but it was no go. They had to give it up. And I tell you, captain, I was just about frozen."

"What did you stop in the boat for all the time? Why did not you go on shore?"

"For the reason that there was three clubs ready to knock me on the head."

"Chipman, why did not you shove [off] and leave them when they was all out of the boat? You could beat the boat up under sail."

"Thought of that, captain, but they took fine good care not to let [me]. They kept me in the stern and they had the boat's painter made fast on shore. One stood by it all the time. Captain, to cut it short, them two rascals intend to steal one of the boats."

"Chipman, I am damn sure they will not if I have to knock in the bottoms." I gave order to have all the oars taken out of the boats and put on top of the officers' shanty and told Mr. Fuller to see that they was all lashed together. When I gave the order them two rascals looked but did not say anything, and it was good for them that they did not at that time. There was some small chain, about ten fathoms with one big link in one end. I had Glass, the cooper, take the end of the chain under the two bow thwarts and rove the small end through the big link, hauling the two boats up snug together. Then I had the end taken around a big cedar log of wood after cutting a score for the chain to lie in. I then had the cooper rivet the chain together. After doing it Glass said, "There, let the damn scoundrels take the boats if they can."

"Yes, Glass, I think I have blocked their little game for this time."

After getting the boat to rights, some of the men helping us — they seemed to know what we was doing it for, for they asked no questions and I did not tell them what I did it for — we had supper. After supper when we had gathered around the fire, we talked the situation over. I asked Mr. Chipman if he thought any of the rest of the crew was in with them two scoundrels. "No, captain, if Reed and Odell could [have] talked over a boat's crew they never would have asked them two, Melrose and Fink, to join them after Melrose had exposed them about they [having] taken the nine sealskins. No, captain, the reason they wanted Melrose and Fink is because they could not make up a boat's crew."

"Well, suppose Melrose had joined them, that would only [have] been four of them and they would want two more men to make up [a] boat's crew."

"Yes, captain, that Reardon and Portuguese Frank would have been very likely the ones to have gone with them."

"Yes, Chip, them is the two; you know I do not trust them two fellows."

"They are just damn fools enough to do what anybody asked them," said Manice. "You see, captain, I was right about Reed and Odell being at the bottom of the boat being over to Atlas Island."

"Yes, Manice, you was right but anyone would have thought they could not have the cheek to ask them for they was the strongest for putting Melrose and his gang on Atlas Island."

"You will find, captain, that Buck Odell has cheek enough for half [a] dozen men," said Manice.

"For the life of me I cannot see what the men are thinking about. Even if they should get to the seal island and get some seal, what are they going to do with them? They ought to know that when [a] vessel comes one word from me will block their game."

"Captain, the most of them thinks that Lawrence & Co. will have a vessel out and they will sell the skins to her captain."

"Well, I am afraid they will be disappointed in their expectations. There will be only one vessel and that vessel will belong to C. A. Williams & Co.; don't you forget it."

"Captain, there is one thing I shall not forget," said Chipman.

"What is that, Chip?"

"Why, them (G—— D——) rascals keeping me in the boat and just about froze me to death, and by the good gods above I will get even with them one of these days."

"They must have planned that rascality after you left for I think none of the crew seemed to know what was going to happen. They was surprised to see the boat over to Atlas Island."

"I think that Odell and Jack Reed has been planning this for some time," said Mr. Fuller.

"What makes you think so, Mr. Fuller?"

"They have been asking me about a good boat crew getting out to the Cloudy Islands. I thought nothing of it at the time, but now I can see the drift of their conversation. You see, captain, ever since Melrose fetched them sealskins up and you got hold of them, they have been down in the mouth."

"I do not see why they should be down in the mouth about them skins. They will get their share of them."

"They expected to get more than their share. When a vessel came you weren't to have nothing to say about the skins," said Mr. Chipman.

"What difference does it make now the skins are up here, Mr. Chipman, I should like to know?"

"Melrose you see knocked their calculation in the head by bringing them up and giving them to you. They intended if one of Lawrence's vessels came out to sell the skins to her, captain, and, you would have nothing [to] say, seeing that the vessel would belong to another firm."

"That is something new to me. How did you find out what they was going to do with the skins?"

"They did not mind me but talked just as if I weren't there. I let them talk to find out what they was up to."

"Well, Chip, I wish that you could have left them on Atlas Island. What fine times they would have had with Melrose."

"Why," said Mr. Fuller, "Reed and Odell would not have been on [the] island forty-eight hours before they would be killing each other."

"That [is] so, Mr. Fuller," said Manice. "Melrose would want to be boss, and if they did not give in to him there would be a fight and that Cape Town Jack would use his knife."

"Yes," I said, "They would find Jack Melrose the best man of the three and [I] wish they was over there now. Even if they did cut each others' throats, they would be off my hands."

"I think, captain, we all echo them sentiments," said Chipman, "And for my part I would be willing to put them two rascals over there with Melrose."

"Well, I should not weep if they was out of the way, but they are not over on Atlas Island, that is the worst of it. We have got to do the best we can with them, but hereafter the two shall not go in the boat together, and I want you officers from this time out, if you have call to hit one of them damn rascals, hit them to lay them out for good."

"I can tell you what, captain, hereafter I shall have something to defend myself with, you can bet your bottom dollar," said Chipman.

It was late when we turned in that night and when I did lie down it was well along in the small hours before I got to sleep, thinking of [the] Chipman episode and how to get the best of them two rascals. [I] made up my mind that night to use the men well, the same as I should like to be used if I was one of the men, and if they did not like that they must do the other thing.

The next day when I came home from my tramp, Chipman and Manice called me [to] one side and showed me what they had been doing that day. They had got out three pieces of lead the shape and size of a hen egg with [a] hole in one end and through the hole they had a lanyard rove through about eighteen inches long with an eye in one end to slip around the wrist. As soon as I saw it I knew what it was. I said, "Why, Chipman, that is one slingshot and it is a terrible weapon to be used by a determined man."

"Yes, captain, and I am that man to use it on them two scoundrels if they give me the chance again."

"I hope they will not give you the opportunity for their own sake for if you should hit one of them [on the] side of the head you would lay them out for good."

"That is just what I want to do, captain, for I have taken all the insults that I am going to from them two devils."

"Well, old man, I do not blame you; only let me have a crack at one of them at the same time."

"Well, captain," said Manice, "What do you think of them?"

"I think they will do when you have nothing else to defend yourself with."

"Well, captain, I should like to get a crack at Cape Town Jack with this. I think I can lay him out."

"No doubt about it, I should not like to get cracked [on the] side of the head with it, Manice."

"You can bet, captain, if you got a hit [on the] side of the head you would have the headache for the rest of the week. Will you have one, captain? It may come in handy some day."

"Yes, Manice, I will take one, for sometimes I leave my stick at the shanty, but I hope I shall never have to use it."

One fine morning I [was] in back where we get our water, sitting down on a rock in the sun, when Odell and Jack passed by me. I said nothing to them. In [a] few minutes Chipman came along, eyes out. [As] he was passing by without seeing me, I said, "What is up, Chipman?"

When I spoke he stopped [and] looked around in surprise. Seeing who it was he said, "That damn Nigger of a Reed told me if I would come in back he would break my head and I am going to give him the opportunity of doing it."

"All right, Chipman, go along; I will be with you in one minute. You play him until I come so I can get a crack at him."

I had to go around a small pond of water and when I came to where Chipman was, on a ridge, I could see that he was disputing with someone. When I came close under him I raised my hand for him not to let on that I was coming. I stopped to hear what Reed had to say to Chipman. He called Chipman all the insulting names he could think of and he said, "You (G—— D——) grey headed old fool, you come down here [and] I will shove this into your damn guts."

I thought it about time to put [a] stop to such talk, so I jumped up alongside of Chipman and said, "Reed, you damn white Negro, you [try to] put a knife into him [and] I will see if [I] cannot put this alongside of your head first." With that I jumped for him. As soon as he saw who was speaking to him, he turned about and started on the run and I after him. I called after him, "Stop and I will give you [a] chance to shove that knife into me, you damn rascal!" It was of no use, he would not stop. Neither could I overtake him. As he was running he was calling out for Odell, saying, "Come and help me, Buck, the captain is here!" When I stopped running he stopped too but he kept calling for Odell. I told him he had better call Odell, "And let him or you come within my reach," holding up what I had in my hand, "And I will have no more trouble with one of you damn rascals."

When I stopped running, Reed stopped too but he kept off to [a] distance. He said, "Captain you have no reason to interfere. I have said and done nothing to you."

"No, I suppose you think that you can say and do what you damn please, the same as you and Odell did when I sent Chipman after blubber to burn. You two forced him to go to Atlas Island where you tried your best to get Melrose to join you two to steal our best boat, but you found Jack Melrose not [such a] big damn scoundrel as you two, was he? No, I have cause to have hard feelings against you."

"Captain, we did not intend to steal the boat but we was going to ask you for her."

"If you was going to ask for the boat, why did you make Chipman go over to Atlas Island that afternoon?"

"We wanted [to] see if Jack Melrose would go with us before we asked you for the boat."

"Cape Town, I think you are a damn liar for you know when

you say that you was going to ask for the boat that you are lying. And another thing: you know that I would not let you have the boat." By that time Buck Odell came back to where Reed was. I said to him, "Odell, you ought to be ashamed of yourself, but you came from a family that has no shame."

"Well, Captain Fuller," said Odell, "Pop Chipman has been telling us all along that he has as much right to the boats as you have. If he has [a] right to the boats I don't see why we haven't as much right to them as he has."

"Well, I want you to understand this much, that Chipman has not as much to say about the boats as I have or you either and I think you lie when you say he says he has. Remember one thing more: I am answerable for you all and for all your doings. No one will look to Chipman or to any of the crew but it will fall on my shoulders. Therefore, you all have got to do what I say or somebody has got [to] go under."

"Captain Fuller, if Chipman would mind his business and talk less to us fellows [about] what we can do we all would be better off."

"Well, Odell, did Chipman tell Jack Reed to come in back and [say] he would break his damn head for him?"

"I know, captain, [that] Jack had no business to tell him that, but Jack was mad to think that Pop Chipman should tell such damn lies about us when he was as much to blame about the boat going over to Melrose as we was."

"If Mr. Chipman was to blame, how was it you kept him in the boat while you was trying to persuade Melrose and Fink to join you two in taking the boat?"

"It was Melrose that kept him in the boat; he was afraid of Melrose."

"Odell, you are [a] damn liar; you know that it was that Cape Town Negro and yourself. Why do you try and lay it on Melrose?"

"That is a fact, captain; you can ask Jack."

"Yes, Cape Town Jack is [a] nice fellow to ask. He would say anything to hurt Chipman. But I shall find out one of these days who is the liar. Now, let me give you [a] little advice. You two had better look out or you will get broken heads for I shall not stand your nonsense no longer." With that I turned to speak to Chipman. I found Manice standing [a] little ways from me. "Where did you come from, Manice?"

"I was down on the point seeing if I could get some mussels for a stew for dinner when I heard Cape Town Jack calling for Odell, so I got on [the] rise to see what was up and I saw you and Chipman running after that beachcomber. I thought there was something up, so I let out to see if I could have [a] hand in the fight if there was going to be one."

"You would have been too late; all would have been over at the fair, Manice, before you got there." We three walked back to the shanty where we found Mr. Fuller and Glass and some of the men. I told them what had happened. They did not say much but I could see that they all kept up devil of thinking.

I had been in the habit of going from the shanty about three miles where there was [a] cave what I called my lookout where I spent the day walking up and down. From this cave I could see down the bay. When I got tired of walking I would sit down and watch and wish that [I might] see a vessel coming from home. After giving them the latest news at the shanty, I went into the shanty to get my walking stick. When I came out Glass said, "Captain, where be you going?"

Said [I], "Where be I going? Why I am down to my lookout; why?"

"I think you are running great risk by going down there alone after what has taken place today."

"Why, what can happen to me when I have this friend with me?" holding up my walking stick.

"That is all right enough, captain, but you know when you are going along to the cave you have to go under the cliff and it is quite easy for them to drop a rock on your head."

"I know all that Glass, but hereafter I shall have a look along the bank before I go under the cliff."

"That is all right about going in, but the coming out is the risk: there is but one way to go and come."

"Well, I am going down there; I cannot stop up here to the shanty all day. It would just about kill me."

"Why, captain, if you are bound to go, hold on one minute until I get my walking stick and I will go down with you," said Glass. In [a] few minutes he came and as we went along we talked the situation over. Glass thought we was going to have trouble with them two rascals. I told him that I had no fears as I thought I had nipped this last trouble in the bud. As we went long we kept a good lookout and kept from under all cliffs. When we got down to the cave, Odell and Reed was there. As soon as they saw Glass, they got up and left saying not one word either to Glass or me. It was evident they expected that I would come down alone as I had been accustomed to doing. After they had left, Glass said, "There, old man, you see them two rascals was waiting for you."

"Yes, they was waiting here in the cave, but whether they was waiting for me is something we cannot tell."

"It is evident enough they was waiting for you: they never came here before today."

"Well, Glass, have it your own way; we will not argue the point." We stopped down to the cave all day and when we felt hungry we catched two rabbits and we broiled them over a fire made of dry peat coals, we having plenty [of] salt. We stopped until about sunset when we returned to the shanty. That night I told the officers that they had better not go away from the shanty unless they went in pairs.

Sometime [about] the first of August, I asked Mr. Fuller to take [a] boat and boat's crew and go down to Pot Harbor and see what Melrose had taken of the stuff that we had left there belonging to Lawrence & Co. and see what they had destroyed, as I wanted to give an account of the stuff to Mr. Williams. And all the stuff that Melrose and his gang has taken and destroyed they shall pay for it. He wanted to know what boat he had better take. I told him that he could please himself, but I thought he had better take his own boat's crew, only taking Carroll in place of Fink. But he did not get away until the middle of the month.

In the meantime Glass and the officers played [a] joke on Manice. Two or three days before Mr. Fuller went to Pot Harbor, Glass was complaining about having a lame back. I told Glass he had better take some pitch and slush[34] together. He [Manice] wanted [to] know what he was making the plaster for. Glass said, "For my lame back."

"Well, Glass, I have [a] lame back too; I wish you would make me one."

"All right, Manice, I will make you one," said Glass.

"But, Jim, I want you to make it big enough to cover my whole back."

"I will make it big enough for you, don't you be afraid," said Glass. So Glass took [a] piece of canvas about two feet long and about one foot and [a] half wide. Then he took some pitch and slush. After boiling it, he put it on the canvas and put it on [Manice's] back. When Glass was putting the plaster on, Manice said, "That is just the thing, Jim."

Well, it happened that Manice belonged to Mr. Fuller's boat's and three days [later] Mr. Fuller started for Pot Harbor, the morning being fine with [a] light northwest breeze. As soon as he got afloat he set the boat's sail and the men on the beach gave him three cheers and off he went in fine style.

They afterwards told me that when they got to Green Island the wind dropped off and it became calm and from there they had to pull. When they arrived at Pot Harbor the pitch plaster had worked under [Manice's] hams and there it stuck across the hind part of his legs, and when he tried to get out of the boat he could not get off his thwart. They had to take his pants and shirt off and haul the plaster off and as they did the plaster fetched skin and all. Manice said, "I don't want no more of Jim Glass's pitch plasters!"

One night as we was sitting around the fire Glass said, "Captain, I think we had better knock [a] hole in [the] bottom of the boat and I think we had better do it tonight."

I said, "Why tonight, Jim, and why knock [a] hole in her?"

"I saw Odell and Reed looking her over and they seemed to be planning something and this is [a] good night on account of it blowing so heavy. There will be no one out looking around and the wind will carry away the sound from the shanty."

"But, Jim, I do not like the idea of putting [a] hole in her bottom and she is our best boat, too; what do you think, Chipman?"

"I would rather knock in her whole bottom rather than let them two rascals have her," said Mr. Chipman.

"But you must look at it in another light. If anything should happen to the other boat, what shall we do for another boat?"

"Why, captain, I can repair her again," said Jim.

"And do you know, captain," said Chipman, "I think them two rascals intend to cut up some rascality while Mr. Fuller is at Pot Harbor. You know they knew that Mr. Fuller was going to Pot Harbor before they forced me to go over to Melrose's shanty; so putting two and two together, they are up to some mischief."

"Melrose would have nothing [to] do with them, Mr. Chipman, and they have got to have more than themselves to man the boat."

"That is all right, captain; perhaps they have talked some of these men over to go with them. You can not tell nothing about such fellows. They may tell you one thing and do another."

"Well, I do not see what men would go with them unless it would be Tim Reardon and Portuguese Frank."

"Captain, you can make up your mind that them two fellows are up to something. You see, Reed or Odell neither of them offered to go to Pot Harbor with Mr. Fuller and generally they have been the first ones to offer their service whenever the boat was going."

"That is true, Mr. Chipman; I noticed that."

After arguing the pro and con, I gave in and told Jim Glass to knock [a] hole in her bottom. "Jim, you get your tools that you will want; the cook and steward and myself will help you, and Chipman, you will keep watch of the men's shanty from around the corner of the cook house and if any of the men come out let us know." When we got outside it was blowing [a] fearful gale of wind. The moon was well up in the heavens, shining brightly — only flying clouds would obscure it from sight every few minutes — but we four did not stop to contemplate the moon; we had other work to do. We turned the boat over on her side so that we could make two holes in her garboard strake. We made two holes on each side of the keel. After putting the two holes in, we let her come back on her bottom again and anyone looking at her would think she was all right. We hardly spoke a word while we was at work on the boat and as soon as the job was finished we made [a] beeline for the house. When we got in, Jim Glass said, "There, them fellows will not get far in her before they will want to bail fast to keep her free. And Mr. Chipman, did we make much noise?"

"No, from where I was standing I could hear not [a] sound and you did the job in quick time."

The old cook spoke up and said, "And faith, mon, did you expect we was going to be all night, mon, when we four are the best men in the whole crew," he being the poorest of the whole lot. The next morning I got up early and went to the boat to see if I could detect where we had made the holes in her by daylight. I found that I could not see the least sign of anyone troubling her. When we was getting breakfast I told Chipman and Glass that them two rascals would think the boat was all right for I had [a] look at her before anyone was up.

"I hope they will not find it out until they get well off into the middle of the bay," said Chipman.

"I should not like to see any of them get drowned, as big damn rascals as they are, Mr. Chipman."

"Well it would not make [a] particle of difference to [the] men and I would not turn my hand one way or the other to save them."

"Well, Mr. Chipman, you think so now, but if you did [see] one of them drowning you would help to save their lives as quick as anyone."

"I might do so, but, captain, if you had been treated by them as I have been I do not think you would hold out your hand to them if you was seeing them going down for the last time; that is how I feel towards them devils."

"I don't know, Chip, but [I] think no matter how hard I might feel towards them two I could not see them drown right before my eyes without trying to save them, no matter what they had done to me."

"You and me, captain, are different in our make up."

"Well, Chip, you are only talking to hear yourself talk."

Mr. Fuller had been gone about fourteen days and we was beginning to look for him the first fine day. The men had behaved themselves very well, when one morning right after breakfast I was standing in [the] doorway of the officers' shanty when all of the men came out of their shanty. Whether they had it so arranged or it was by accident is more than I can tell, but I think it was more by accident with the most of them. Odell came up to me; I noticed that Reed kept close alongside of him. Odell said, "Captain, can we have the boat to go to Swain's Islands?"

"No, Odell, you cannot have the boat to go to Swain's Islands without an officer goes with her."

"What in hell is the reason we cannot have the boat," said Jack Reed, and with that he struck me a blow in the face, which I partly warded off, and I shot out my right and took him under the nose. That staggered him. I had got him by the throat when I saw him reaching for his knife. I heard Glass say, "Look out, captain, he is after his knife! Let me get a crack at the damn rascal!" As I jumped to one side I heard some of the men say, "None of that knife business, Cape Town Jack." But he did not stop to say anything, but tried to keep clear of Jim Glass's stick. Glass made [a] strike at him, but somehow he dodged the stick. I saw that Jim could not overtake him, so I said, "Let me have the stick, Jim." With that I grabbed it out of his hand and chased Reed around the shanty. Chipman grabbed him once, but, before I could get to him, he broke away from him. When Chipman caught him, he bellowed like [a] bull. After he broke loose from Chipman he kept clear of the men that tried to intercept him. To get clear he dodged into the men's shanty and shut the door. Chipman went to open the door to dislodge him. As he opened the door a brick came flying at his head, but did not hit him. I said, "How be we going to get that devil out, Chipman?"

Before Chipman could speak, Glass said, "Chipman, shut the door until I can get the keg of gunpowder and I will put a fuse to it and blow the shanty to Hell and him with it." With that Glass got his club hammer and was knocking the head in the cask where we kept the powder when he happened to look up and he saw Reed going over the hill. He came to us and said, "You are damn smart ones to look after that rascal; why he has got out and gone over the hill as if the devil was after him."

"Well, Jim, he must have heard you say that you would blow him to Hell with the shanty." Chipman opened the door. True enough he was gone and [a] large slit in the canvas showed the way he got out. "Ho, Jim, I wished he had waited until you had got that keg of gunpowder out of the cask; then he would have thought the devil was after him for sure." After seeing that he was out of my reach I called all of the men and told them that [they] must find that beachcomber and tell him that if he comes back to the shanty and I or any of the officers catch him we will smash his head for him. "You tell him that he will get nothing to eat from the shanty, and if I find any of you men giving him provisions I will turn him away from the shanty and he can go and live with that rascal. And Odell, he is your friend; I want you to impress it on his mind that he had better keep away from the shanty until there is a vessel comes. If he does [come], as true as there is a God above, there will be one less to go away when the vessel does come."

That afternoon Odell and one or two more men went and looked for that scoundrel, but when they came back they said that he could not be found. That night after supper as we was sitting around the fire, I told Chipman [and] Glass that we must keep a watch at night after this for we could not tell what that beachcomber might do. "And Mr. Chipman, them whale lances, I have put them overboard but the three elephant lances are on the back side of the shanty, and I think we had better bring them into the shanty for safe keeping. After bringing in the lances, Mr. Chipman said, "Captain, I told you sometime ago that them two devils was up to some mischief and I thought it would break out while Mr. Fuller was away."

"Yes, I know you did, and your words have come true."

That night, Chipman, Glass and myself kept watch, each standing as near three hours as we could think, we having no time piece to go by; but the night passed by without anything happening, as usual we hearing or seeing nothing of Jack Reed, it being

[a] bright moon light night — in fact, almost as bright as day. The next morning the wind was blowing strong until about sunset, when it began to drop off almost a calm. I told the men that I expected Mr. Fuller the next day if it kept calm. The men seemed to be anxious for Mr. Fuller's return. They wanted to know what I was going to do about Cape Town Jack; if I was going to let him wander around just as he pleased and do what he had a mind to do. I told them that I thought we had better wait until the other boat's crew came and see what they thought about it before we decided what to [do] with that beachcomber. Afterwards Mr. Chipman wanted to know why I told them that I should wait until Mr. Fuller came before I decided what I was going to do with Jack.

"Why, Chip, don't you see the point, why some of them fellows is in with that Cape Town Negro and all we say will be repeated to him."

"I see, captain, you are right, there is that Odell for one."

"Yes, Chip, and there is others besides Odell; don't you run away with the idea that they have not seen Cape Town Jack."

"Why, when did they see him do you think, captain?"

"They saw him yesterday; that was all bosh about not seeing him and if the truth was known some of them fellows have seen him today, I am thinking."

"Why, who do you think the men are that is in with him, captain?"

"You stake your life for [a] chew of tobacco that Odell and Portuguese Frank and Tim Reardon are the three rascals. If they weren't in with him, why did they ask me what I intend to [do] with him, I should like to know? The other men don't seem to care what becomes of him."

The men seemed to be sorry for what had happened. Even Odell seemed to be taken aback by what had taken place. He stopped around the shanty all day with the rest of the men. Neither did the officers go from the shanty. It seemed as if we was living over an active volcano and we did not know when it would break out. That day Glass and myself took the keg of gunpowder out of the cask and dug [a] hole in the ground in one corner of our shanty about three feet deep and we put the keg of powder in the hole and covered it up with sand. Chipman [and] Glass thought it was [a] risky piece of business putting [it] in the shanty. "Not half as risky as leaving it out there in the cask, Chipman; you know that fellow could roll that cask in back some night and take out the powder and blow us all to that hot place."

"Well, if [a] spark of fire should get to it now we will go to kingdom come, that is certain," said Jim Glass.

"Let her blow up, Jim, we shall see who will get to kingdom come first, but don't you be frightened. It is not going to blow up, for how can fire get to it when there is two feet of wet sand on top of it?"

That night while we was talking over the situation, Mr. Chipman said, "Captain, how would it do for us, when Mr. Fuller comes from Pot Harbor, for us in the officers' shanty to take the two boats and go to Butler's Harbor[35] and leave the rest to get along the best they can. I think we have put up with enough from the whole lot to serve them out some way."

"No, Chipman, we cannot do that. In the first place, [we have] no stuff to put up another shanty with; the second is how be we going to get our provisions down there without they knowing of it; and lastly, Chip, some of the men have tried to do what is right and they have stood by us in everything and I could not desert them. No, we will stick to the old shanty as long as we can and I think it will be until a vessel comes from home."

Three days after the trouble with Cape Town Jack the wind was light from eastward, when well along in the afternoon we saw the boat returning from Pot Harbor. As soon as she was seen everyone was out looking and all seemed to be as anxious as if she was direct from home, but the men in the boat did not seem to be in no great hurry. They had the sail set but the wind was so light that the boat did not make much headway, and it was about sunset when she got to the landing. They had brought part of [a] load of blubber and some wood and it was about dark when [we] got the boat unloaded and hauled up on the beach. And when I told them about Cape Town Jack, they did not seem much surprised at what had happened.

While we was getting supper, I asked Mr. Fuller what stuff

Melrose had taken. He said Melrose had taken the head out of the bread cask and all of the casks that had sails in and by doing so had spoilt about all the stuff down there; even things that he did not want he opened and left open.

"I tell you what, Captain Fuller, it was damn lonesome down there," said Manice.

"Yes," said Mr. Fuller, "one night after we had been there about [a] week, Manice began crying like [a] little baby. I asked him what he was crying about. He said between his sobs, 'Mr. Fuller, suppose a vessel should come into Norton's Harbor while we are down here and would not wait or come for us. It would be damn hard,' and he bohoed the harder. I said, 'Manice, you are [a] damn big fool. Don't you know the old man knows that we are down here to Pot Harbor and you need not be afraid; there will be no vessel out before we get to Norton's Harbor.'"

I said, "How was it, Manice? I thought you had more backbone than that to give away to your feelings before the boat's crew."

"I could not help crying. I was so low spirited that night and I got to thinking if such [a] thing should happen what would become of us poor devils?"

"You ought not [to] let such thoughts run in your head and if you did want to have a good cry, why you ought to have gone outside and no one would have been any the wiser."

"It pleased the Portuguese to see [him] crying. They made all manner of fun of him and when [he] got mad the Portuguese had [a] good laugh at him," said Mr. Fuller.

"I don't blame the Portuguese for laughing at him. I would have the same myself if I had been there."

"And mon, so would I," said the old cook.

"Yes, and, you old devil, you would have got a punch in the nose," said Manice.

"And, mon, I don na about that for that is something two can play at, ma biy."

That was too much for us in the shanty to think of, the old cook holding up his dukes to a man like Manice. We had to laugh — not only laugh, but roar!

That night we kept watch. Somewhere about midnight Mr. Chipman had the watch. He called me [saying], "Captain, I think Cape Town Jack is around but it is dark as thunder outside. I cannot see no distance."

"What makes you think that Jack is around?"

"I heard somebody in the cook house."

[I] got up and had [a] look outside. Sure enough it was dark as my pocket; it was half rain and fog. I said, "Chip, you had better not go outside of the shanty. You may get a brick alongside [of] the head from that rascal."

"No, captain, I shall not give him the chance to hit me in the head with a brickbat in the dark."

"I think you had better light the lamp and put it in a dark corner."

While he was [at] work lighting the lamp I slipped outside, but everything was still — not a sound could be heard, and when I came in, I said, "Chipman, you must have been mistaken."

"No, captain, there was someone in the cook house when I called you."

The next morning we found that we had had a visitor in the night, for we found about half of the canvas gone from the roof of the cook house and he had taken some bread and pork. After eating breakfast we patched it over with some spare canvas and told Odell and Tim Reardon that I wanted them to take Cape Town Jack's blanket to him and tell him that he came near getting into trouble last night and he had better not try it too often.

"I think you will find him in the cave where Mr. Fuller found the sealskins."

They went off and was gone all day and when they came they said they could not find him but he had been in the cave for the pork was hanging up and the bread was in some canvas rolled up, so [they] left the blanket in the cave. I did not believe them when they said they did not see Cape Town Jack. I said, "Odell, what made you so long? It did not take you all day to walk over there and back, did it?"

"We did not hurry and we stopped sometime in the cave waiting to see if Jack would come back."

"What did Jack have in the cave? Did he have any fire wood?"

"Yes, sir, he had some blubber and [a] large bundle of Desolation tea roots in the cave."

"I suppose you did not make a fire, did you?"

"Oh, yes, sir, we did and cooked something to eat."

Afterwards, talking it over with the officers, we came to the conclusion that them two fellows had seen Jack Reed and had given him all the news.

Two or three days after that I was walking up and down on the sand close to the water where the sand was hard, thinking of home and wondering whether there was a vessel on her way looking for us when Carroll and Magill came along. I asked where they was going. They said, "We thought of going to see if we could get some rabbits."

I said, "Well, boys, I should like to go along with you but I can not for there is no one at the shanty except old man Glass and it won't do to leave the old man alone at the shanty."

"Captain, it is mighty hard that that damn Negro should be at large. Why [don't] we all get together and hunt him down and put him on one of the islands?"

"Why, Carroll, you know why that cannot be done without asking. You know as soon as we make up our minds to do so, why Odell or some of the crew will be off and tell him and then you might as well look for [a] needle in a haystack as to look for him."

"If Odell thinks so much of him, captain, put him there too. Why, I notice you have not been away from the shanty since we got back from Pot Harbor and if I was you I would not stand it all for that damn white Negro."

"Carroll, I don't mind letting you two [know] what I have been thinking of doing one of these days. I have [been] thinking of going down to Butler's Harbor with the officers and stopping there until there is a vessel comes from home."

"Oh, captain, don't leave us behind; take us with you for God's sake when you go."

"I shall not go unless I have more trouble with the men. I have made up my mind that hereafter we shall live in peace. If not I shall go there and leave the rest of the crew to look after themselves."

"Captain Fuller, we cannot blame you for wanting to go, for you have put up with a good deal from first one then the other."

"I should have left you men long ago but I knew there would have been some bad times for them that was left behind with Jack Melrose, Odell, and Cape Town Jack — them three would want to be boss."

"It is a good thing that Jack Melrose cut up when he did," said Magill, "and now he is boss of all he surveys with his man Friday."

"You may say what you have a mind to about Melrose; I will bet that he is happier than we are. You see, when he came over and took them nine sealskins he done it for [the] price [of] devilry and to let Cape Town and Odell see that he did not care a curse for them."

"Yes, sir, and weren't Jack and Odell mad when they found the skins gone? They would have killed Melrose if they could have got hold of him at the time," said Magill.

"That is so, Magill, but Cape Town and Odell would have found when they got hold of Melrose that they had not Flaherty to deal with. I would not have been afraid if I had been a betting man that Melrose would have given them two broken heads."

"Why, captain, they was afraid of Melrose," said Carroll, "And they was glad when you put him on Atlas Island. But, captain, if you should go to Butler's Harbor, will you leave the old boat behind for the crew?"

"No, Carroll, I will take both boats with me."

"Well, you will find that most of the men will want to go with you; I know I shall," said Carroll.

"And me too," said Magill. With that they left me and I thought to myself, "Well, I have given you something to think over, my boys."

I then went to where Jim Glass was and told him what I had told Carroll and Jim Magill. He said, "What did you tell them that for? They will tell the rest of the men."

"That is just what I want them to do."

"Yes, and after this they will keep an eye on you and you will hardly move without someone looking after you."

"That will divide them up, Jim, don't you see, and we shall live in peace for a while."

"Captain, I think you have made a mistake in telling them what you thought of doing."

"Do you think so, Jim? Perhaps I have, but I think you will see that I am right, old man."

For the next three or four days things went along smooth as clockwork, no quarreling or fighting. When that joyful day, the eleventh day of September, arrived, a day that will never fade from the memory of the crew of schooner *Pilot's Bride.* Oh, well do I remember that morning. There was strong wind blowing at daylight, but clear overhead and along the tops of the mountains to the northwest was a heavy bank of clouds shutting out the tops from view. But as the sun rose higher and higher in the heavens the bank began to thin out and [the] wind to moderate, and when noon came it was blowing a good wholesale breeze. It was something strange that morning. Everybody had stopped at the shanty, not one thought of having a look to see if there was a vessel in sight, something that had not happened before that day. No one had been further than the spring after water. And I remember hearing the old cook saying when dinnertime came, "Well, you fellows are all at home today for a wonder."

And Tom Flaherty saying, "Yes, cook, let's have dinner."

"And faith you had better let the officers get their dinner first, Flaherty; if not the officers, let me wait on the Captain first."

I heard the whole of the conversation as I was coming out and I stepped into the cook house and said, "Never mind the officers or the captain, cook; wait upon the men first."

Carroll and some of the rest spoke up, "Not by a damn sight, cook; give the captain and the officers their dinner first."

"And faith, mon, them that cannot wait for their dinner until I can serve the captain and officers, they can cook their own dinner. Herafter I will neither cook or wait on them and I want [them] to mind that now."

I said, "Cook, never mind them fellows."

"And faith I do mind them," said the old cook, "and by my faith, captain, how was them fellows brought up, I should like to know?"

"I will tell you, cook," said Carroll. "You see some of them was brought up in a pigpen."

"And faith you are about right there, Carroll."

The cook after growling a little while he gave us our dinner. Someway Mr. Fuller finished eating first. He got up and said, "Well, I am going up on the hill and see if I cannot raise a vessel."

As he was going out of the door, I said, "You will not see no vessel until about this time next month, Mr. Fuller." He said nothing but shut the door and walked away. The rest of us kept on eating.

Mr. Fuller had not been gone more than two minutes when he sang out. "Oh," I said, "That must be Melrose, but I will have a look." I got up and went to the door and poked out my head and looked down the bay. "Do my eyes deceive me? Is that a vessel's flying jib boom coming around the point? God above, boys, it's a sail, a sail!"

The words had hardly passed my lips when it was taken up by the men. Away went rabbit stew and we rushed from the shanties like wild men, swinging our hats and shouting, "Sail ho!" We all went for the boat, some getting the oars while the others was shoving down the boat and all talking at once. And I think if our friends had seen us at that time they would have said why they have all gone crazy, and they would have been about right, we was crazy, crazy with joy, for we knew that we was going to hear from home, from wife, mother, father, sister, and brother, from those that we had not heard from for over two long years. We did not know whether it would be good news or bad, but we was willing to take our chances only to hear from home. When the boat was in the water, everybody wanted to go. I said, "Men, I am sorry to disappoint you, but I want only a boat's crew. You see she is beating and we can take only one boat on account of the boat's being in the way of her working, but as soon as she comes to anchor I will send the boat after you all." With that they said, "All right, captain."

With that I got into the boat and we shoved off and as we was going down the bay we was speculating what vessel it was, Glass saying she looks like the *Francis Allyn*, and Chipman [saying] she looks too large for the *Francis Allyn* and too new. I said, "You cannot tell by that, Chip, putty and paint will make an old vessel look new. It is the *Francis Allyn;* there is Bob Glass on the house."

The schooner had dead beat off it on account of the wind blowing direct out, so we kept well over under Norton's Island so as to board her when she came around. As the schooner came up into the wind we shot alongside and I was on the house and had Captain Glass by the hand, but as for speaking it was more than I could do. There was a big lump that would keep coming up in my throat whenever I was going to speak. There was Mr. Usher, the mate; Mr. John Glass, the second mate; and an old steward who had been with [me] for five voyages. All I could do was take them by the hand. But poor old man Chipman, he broke down completely and had [a] good cry. He turned around to me and said, "I can not help it, captain." But I could not answer him for if I had tried to speak one word I should have been blubbering like [a] school boy.

After [a] few minutes I got my feelings under control. I went in asking and answering questions, when I happened to look forward and behold, there was Cape Town Jack. "How in the name of God did that damn scoundrel get on board of you, Captain Glass?"

"You mean that fellow forward? Why blast him, as I was beating up, Mr. Glass raised smoke and I thought it might be you, so I sent in the boat and when they came off they had that blamed rascal. As soon as I saw him I told Mr. Usher that you had had trouble with him. I asked him what he made that fire for. He said so that he could get on board to let me know where the seal are. I told him, 'I don't care where the seal are. I did not come after the likes of you, you blamed scoundrel, I came after Captain Fuller. Where is he and the rest of the officers and crew?' He said they are all up to Norton's Harbor except two over to Atlas Island and I asked him what they was doing over there. He said they had stolen [a] boat and when they came back you put them there for safe keep-

ing. And [I] told him if the boat had not been hoisted up I would
put him on shore again. 'What have you been doing that you are
not with the rest of [the] crew?' 'Captain Fuller was going to split
my head open and he would not let me stop with the rest of the
crew. He sent word to me that if I came to the shanty again before
[a] vessel came he would lay me out for good.' And I told him that
you must have good reason for driving him away from the rest of
the crew."

"Captain Glass, I should like to have you put that damn rascal
on shore again and keep him there until you are ready for home,
for he has kicked up Hell long enough. I suppose you saw my two
prisoners on Atlas Island?"

"Yes. When I tacked they was waving their hats. That fellow
said you had put them there for stealing a boat."

"Yes, Captain Glass."

Mr. Usher said, "Captain, that pimp of yours said the seal is
on Cloudy Islands."

"Yes, Usher, that [is] where the seal are."

"Captain Fuller, how did you get on shore on the Rocks of Despair," said Captain Glass, "and when was the date?"

"I went on shore the second day of October in [a] thick
snowstorm."

By that time the schooner had got up to the anchorage and as
soon as the anchor was dropped, I sent the boat in for the rest of
the crew that wanted to come on board and I had that Reed put on
shore. After coming to anchor [Captain Glass] gave me two letters, one from my wife and one from Mr. Williams. I think his letter was dated from Paris, but I did not mind his letter; it was the
letter from home that I was most anxious to read. It contained
some good and some bad news.

After I had finished reading my letters and was thinking about
what they contained, Mr. Chipman said, "Captain Fuller, my
wife writes me that she had a letter from your wife in April, saying that she heard from a spiritualist [that] the schooner *Pilot's
Bride* was lost but the crew was on an island living in a canvas
house and she thinks we are all right only we have lost the vessel."
I made some laughing remark about spiritualists and thought

nothing more about it. But if I had heard the whole story I should have been [a] little more inquisitive at the time to think that a man could in his sleep tell of another man who was twelve thousand miles away being shipwrecked and that man being a total stranger to him and to his family, they never before hearing of his name. It was the most marvelous and incredible thing, he not only telling of the loss of the vessel but he told each officer's disposition and described our features and told [the] name of the vessel and captain that was coming to look us up about a month before she sailed or thought of sailing. I will tell the story as it was told to me by my wife and substantiated by my brother Richard. Perhaps some scientific person can explain the thing, but it is more [than] I can do. But I can vouch for what he told my brother. [It] was the truth except one thing, that was he said the vessel caught a fire and still he might be right for all we know. This is my wife's story.

"One day about the first of April at noon, one of my neighbors' daughters sang out from their yard, 'Mrs. Fuller, they have heard from Captain Fuller, have they not?' 'No, Anna, not as I know of.' 'Why, yes, they have, and it's all over town that he is shipwrecked and is living on an island and they say that Richard Fuller up to the Center has heard from him.' 'Why, Anna, it is very strange that Richard should hear from him and I not but I will soon find out.' What that I went to Mrs. Fowles, another neighbor and I said, 'Mrs. Fowles, they say that Joseph's folks at the Center has heard from him and he has been shipwrecked and is living on an island. I should like to have you mind the children while I go up to the Center and find out. Will you be so kind as to do so?' 'Why yes, Mrs. Fuller, and I only hope they have heard from him and I will come right away over.'

"I got ready and as soon as Mrs. Fowles came I took a hack and went to the Center and so happened to find Richard at home. And I said, 'Richard, have you heard from Joseph?' He said, 'No, Jennie, I have not.' 'Why, Richard, what is that story going around town saying that you have heard from him and that he is shipwrecked and is living on an island?' 'Why, Jennie, I think I know what you mean and I will tell you all about it. I was down to Mrs. Cross's last evening. They, you know, are spiritualists. They asked me to come down as George Knight, the great spiritualist medium from Salem, was going to give some sittings. Jennie, after he got to sleep he turned to me and said, 'You and your people are worried about your two brothers. They are shipwrecked and are living on an island. The ship looks as if she had been [on] fire. But everybody is on shore living in two canvas houses. There is little chief and one old greyhead man next to Chief Moses and Jim, big dark man. Little chief short dark man; the men worry little chief. Little chief takes off his coat. He God-damns the men. The men keeps clear of chief. And the chief worry too much about the men but he will come out all right with the men; most of them likes little chief. Moses is happy-go-lucky fellow, whistling and singing all the time. He looks some like the chief, but bigger and taller. Jim he great big man; he is good man, great friend of chief; and there is old man with white hair next to chief. The men do not like him. He makes plenty of trouble for the chief with the men but he does not mean to make trouble but he cannot help it; he talks too much with the men and tells them what chief going to do. And there is a little woman she lives on a hill with three pappoose, wife of chief. She says all the time, 'My poor Joseph. Why don't you come?' You tell her Joseph little chief has lost his ship and is living on an island and he is well. And tell her in May they are going to send a vessel after chief and the men and the name of chief that is going to look for him is Bob Glass; he is big dark man. And you tell little chief's wife she must stop crying, and she will hear from chief the fourteenth of February next year and see him in April.' And Jennie, now you know as much about Joe as I, but I do not put much faith in what he said; but still it seemed very strange that he should [have] picked on me the first thing and said that your two brothers was shipwrecked, for we have not given him up yet you know.' 'Oh, Richard, I did hope that you had heard from him,' and I had to sit down and have a good cry. Richard tried to cheer me up. He said, 'Perhaps, Jennie, God has taken this way through George Knight to let you know what has happened to Joe and that he is all right. At any rate I shall think so until I hear different. And do you know I feel as if he had told me the truth about Joe.'"

And my wife said to me after [she] told the story about the

spiritualist, "Do you know, Joseph, that after Richard had told me what the spiritualist had said and I had a good cry over it, I felt better than I had for [a] long time and I went home to our little ones in better spirits, feeling that you was all right and in time I should hear from you. You, my dear husband, will never know how long the time did seem to me before I got the telegram from Mr. Williams saying you was found and that you was well. I felt as if I must tell someone or I should die. I picked up little Gertrude and ran over to Aunty Fowles with [the] telegram in my hand. She saw me coming like a wild woman and she met me at the door and took little Gertrude out of my arms and said, 'Jennie Fuller, sit down in that chair and don't you speak.' I dropped into the chair but I could not speak if I wanted to or if she would have let me and she took the telegram out of my hand and read it. While she was reading it I had a cry and then she had a cry to keep me company and I tell you, Joseph, she is a good woman."

After we came to anchor and I got through reading my letters, Captain Glass told us about the loss of the bark *Trinity* and how the owners of her could not or would not send a vessel to look for them, so Captain Williams had to stop at Heard Island from October 1880 to February 1882 before he got away from Heard Island, and then the government had to send the man-o'-war *Marion* after him.[36] I asked Captain Glass if the man-o'-war stopped at any place in Kerguelen. He said, "Yes, she stopped one night in Greenland Bay and then tried to get to Three Island Harbor. Someway they could not or did not want to get there, I do not know which; at any rate they did not get there."

"If the man-o'-war had got there they would have found a letter there from me, stating what happened to me and where they would find me, but, Glass, I should have thought that Captain Williams would have told the captain of the man-o'-war that there must have been something happen to the *Pilot's Bride* or I should have been over to Heard Island long before the man-o'-war got to Heard Island."

"Captain Williams said that he did speak of you to the captain, but the captain said the government only sent him to look for the bark *Trinity*'s crew, not for the *Pilot's Bride*."

"Well, Captain Glass, it was damn hard lines to think that

there was [a] vessel within twenty miles of us and the captain almost knew that there was something the matter with us. Instead of looking for us, he left us to our fate and the vessel being a man-o'-war at that, I suppose they was in a hurry to get back to the Cape of Good Hope or to Rio, there being no balls and parties at Heard and Kerguelen islands."

"You said, Captain Fuller, that you had a letter at Three Island Harbor; how did you get it there?"

"I went there in a boat and left it in Fort Independence, and I tell you I would stop a long time down here before I would do the same journey over again."

"It must have been tough work getting back to Pot Harbor," said Captain Glass.

"No, it wasn't the getting back; it was the getting there was the worst part, for when we got ready to come back we left the boat at Malloy Point and walked overland to Pot Harbor. Captain Glass, I want to ask you one thing. When did you hear that I had found the seal island?"

"I never heard nothing about you having sealskins until they thought you was lost, and it was a big surprise to everybody I can tell you. Even Jennie, your wife, knew nothing about it. And when [that] vessel brought your oil home, Captain Smith [and I] got acquainted, but he never let on that you had taken sealskins."

"I asked Captain Smith not to say nothing about the sealskins and he said he would not and I see that he did keep his word."

"Well, Captain Fuller, how many seal do you expect we shall get on the Cloudy Islands?"

"That is more than I can tell. You see we only worked one of the small islands for the seventeen hundred that we got that season. The last day when we took Mr. Fuller off before going to the cape, I sent in all three boats on the upper big island and before they got in to the land it came in thick fog; but they landed in a place what we called Tony's Harbor. They said there was about four hundred old whigs or old male seal."

"If they saw as many whigs as that, there ought to be two or three thousand female and young seal there in the season."

"Yes, I should think there ought to be all of that many there, but I did not have time to give the islands a good look. But, by my reckoning, the islands have not been worked for more than fifty years."

"Well, Captain Fuller, I have salt enough for seven thousand skins and when I get the schooner broke out and ready for work we will come to some agreement with you about what portion you and your crew shall have."

"All right, Captain Glass."

"And, Captain Fuller, your officers and crew will have to sleep on shore until I can make arrangements to take them on board."

So I told the officers what Captain Glass said and they was satisfied and when they got ready to go on shore the old steward came to me and said, "You are not going on shore to sleep, Captain Fuller, are you?"

"No, steward, the soft side of a plank will do for me. I have had enough of sleeping on shore."

"You shall not sleep, captain, on no plank, as long as I have a bed to sleep on and don't you forget it, Captain Fuller."

"All right, steward, I thank you for your kindness."

"You will take my bed, won't you, captain?"

"Steward, if I take your bed where be you going to sleep?"

"Never mind me, you take my bed and I will make out all right."

I saw that if I did not take his bed he would feel bad about it so I said, "Yes, steward and I thank you for it."

"I don't want no thanks; all I want is for you to take my bed and to use it just the same as if it was your own for as long as you're on board of the *Allyn*."

For the next fortnight Captain Glass [was] busy employed breaking [out] and landing stuff and getting ready for work, my crew the most of them helping. By the first of October Captain Glass had the schooner all ready. One day he said, "Captain Fuller, I will tell you what I will do. I have had a talk with my officers. They and myself think that if we give you and your officers and men, that is as many as will go and help to work the islands for seal, one-third of all the sealskins that is taken this season, we finding everything, salt and provisions, and we taking the skins to Cape Town for you. Are you satisfied with them terms?"

"I am, Captain Glass, and I think the officers and men will be satisfied with them terms."

That same day I told Mr. Chipman, Mr. Fuller, Jim Glass, Manice, cook, steward, and Odell. They was all satisfied with the offer, all but Odell. That afternoon I went on shore and told the men what Captain Glass had offered to do and [asked] if they was willing to take up his offer.

"No," they said, "We want half of the skins that is taken."

"Well, men, I think Captain Glass has made us a good offer, he finding everything, salt and provisions."

"Captain, we don't think so. All of his men is green at the sealing business and we think therefore that we ought to have one-half of the skins that is taken."

"Well, no doubt the men are green, but I want you to understand that all the gang aft is old sealers and they can skin two seal to your one."

"We don't go short of one-half," they said.

By what they said I could see that Odell and Cape Town [Jack] had been talking to the men and had talked them over to their way of thinking so [I] said, "I will give you until tomorrow morning to think [it] over." That night at supper I told Captain Glass and his officers and men what the men wanted. Mr. Chipman said, "I could have told you that before you went to them. It has always been Odell and Cape Town Jack's talk that [they] should have half of the skins no matter what vessel came out, and they have talked the rest of them over."

"Well, them two will have to take less for them two shall not have a share in it now at no consideration."

"They must be [a] foolish lot of men to think that I can do better than that by them," said Captain Glass; "that is what we give at Cape Horn."

"But they are thinking that when I go away in the schooner I am going to leave the boats at Norton's Harbor, but they will find themselves mistaken."

"Why, what be you going to do with them? Captain Glass has four boats now."

"You see, Mr. Chipman, Glass will put two boat's crews on shore and I shall put one and one of Glass's crews will take one of our boats and then he will have three boats to work from the schooner." That same day Glass [and] the cooper went on shore and repaired the boat where we had made a hole in her and he said that when they turned the boat over to get at the hole, Tim Reardon said on seeing the holes in her bottom, "Why, cooper, it's [a] damn shame to put holes in the boat like that, and Jack Melrose ought to have his damn head smashed for doing it."

And I said, "Jim, let him think it was Melrose's doing." I told the officers not [to] let on about me taking the boats when we go away from Norton's Harbor.

"Well, Captain Fuller, I hope you will take the boats if [the men] do not go," said Chipman, "And if we don't [take] the boats we can put them on shore in some other place."

After breakfast the next morning with Captain Glass, about nine o'clock I went on shore. When we landed, all the men was in the shanty except one Portuguese. I said, "Well, Manuel, be you going to take Captain Glass's offer?"

"Me going do what captain want me to do. Me no going to be damn fool and when me get in the cape me have no moneys."

"All right, Manuel, you get your things and put them in the boat." With that I went along to the shanty and I opened the door and I said, "What have you made up your minds to do, men, take Captain Glass's offer or not?"

Ed Carroll was the spokesman. He said, "No, Captain Fuller, we won't take one-third. We want one-half or we want nothing. We do not think Captain Glass is doing what is right when he offers us less than one-half."

"Well, Carroll, you and me think different about the offer. You know Captain Glass has got to make a voyage and to pay his way. Who is going to pay Captain Glass or his crew for coming to look us up?"

"Why, the owners will pay him for coming."

"The owners will pay him! Did the owners know whether we was lost on the island or on the passage home? You must look to both sides of the thing."

"Captain Glass will be well paid for coming to look us up if he gets one-half of the seal there is on the Cloudys."

"What do you know or anybody else know what number of seal there is going to be in the Cloudy Islands? I think Captain Glass has made us a good offer when he could have gone there and left us out altogether. You know that Cape Town Negro gave the seal islands away before we got on board of the *Allyn*. You can thank him there for that. And Carroll, you know this much, that if we or any of us should take the boats and go to the island, what good would it [do] us or them? We have no salt and we have no provisions and how be we to take care of the skins after we get them? And then again how be we to get them away from the island? I think myself take it all in all that Captain [Glass] has made us a good offer, in fact all he could offer us and come out whole himself."

"Captain Fuller, it's of no use talking. We will not go for less than one-half."

I saw it was of no use talking to them, so I turned to Odell and Jack Reed and said, "[As] for you two, I would not have you go under no consideration, for you two are damn scoundrels and if the rest would have gone you two should not."

"I don't care a God damn what you say or think of me, but I will get even with you; remember you are not out of the woods yet, Captain Fuller," said Reed.

"Now, men," I said, "Do you have the same mind as Carroll?"

"Yes, we want one-half," they said.

"Well, men, you won't get no more than one-third and all I can say is this: you are a foolish lot of men to be swayed by one or two men that you know to be rascals. And now, men, I am done with you until Captain Glass is ready to take us to the cape, for I am going to take Glass's offer. There is only one man among you that has any sense and that is Manuel the Portuguese."

Captain Glass said, "Men, if you do not take my offer I shall not find you with provisions and when I leave here to go [to] the seal islands I don't want you to trouble my provisions."

Odell said to him, "If you don't want to find us, why don't you take us to the cape then; you was sent from home to get us."

"I think you are a little out there, my man," said Captain Glass, "And I want you all to understand this much. I wasn't sent to get one of you blamed rascals. I was sent to find Captain Fuller and if I did find him after the season was up I was to take him to the cape. Nothing was said to me about lumping the crew to Cape Town."

"Well, Captain Glass, I think you ought to give us one-half; we found the seal," said Odell, "And I think you damn mean only wanting to give us one-third."

"You had better mind what you say or I will leave you to get away from the island the best way you can for if you are saucy about it not a blame one of you shall set your foot on the *Allyn*'s deck."

"We won't go to the Cloudy Islands; we will get along with the provisions that belong to the *Pilot's Bride*," said the men, "And you, Captain Glass, need not be afraid — we shall not trouble your provisions."

"Very good, men," I said, "[I] shall not ask you again if you alter your mind." With that, Glass and me left them. We took the Portuguese and went on board of the *Allyn*.

That same afternoon I asked Captain Glass if he would let me have three of his men to go over to Atlas Island and see Melrose and Fink and to see what they thought of doing. "Why, yes, you can have a full boat's crew if [you] want them."

"No, captain, three men will do me." So I took the boat and went over there. Melrose and Fink was glad to see me. They was down to the water long before I got there, waiting. As soon as I got there, Melrose said, "How do you do, Captain Fuller?"

I said, "I am all right, Jack; how be you two fellows?"

"Oh, we are all right, captain." They was as polite as you please. All that bombastic was gone out of them. They wanted to know what schooner that was. I told them the *Francis Allyn*, Captain Glass. Then Jack Melrose said, "Did they fetch any letters for me, captain?"

"No, Jack, but I have brought you some papers," and I gave them the papers and I said, "Now, Jack and Fink, I want to know if you two are willing to help work the Cloudy Islands for your lay out of one-third of the seal that is taken on the Cloudy Islands."

"Captain Fuller, how many of the crew is going to help work the islands? Is Cape Town Jack, Odell, Ed Carroll, Jim Magill, Tom Flaherty, Reardon, and all the Portuguese — be they all going?"

"No, Jack, they wanted Captain Glass to give them one-half and Glass told them that he would not give them one skin over one-third, he finding everything — salt and provisions — and he taking the skins to the cape. Well, Jack, they said that they would not go for no less than one-half. Then I told them [they] could stop where they was and be damned to them until the schooner was ready to go to the cape."

"Who is going, Captain Fuller?"

"No one but Mr. Chipman, Mr. Fuller, Manice, cook, steward, and Manuel the Portuguese. Now, Melrose, will you and Fink go?"

"Yes, Captain Fuller, if you will make my lay one hundred and Fink's the same."

"All right, Jack, I was going to offer you a better lay."

"You see, captain, if them other suckers had been going I would rather stop where I am. I don't want nothing to do with them, but, as long as they are not going, I and Fink am with you."

"Well, Jack, if the rest of the men had said that they would have gone, Odell and Cape Town Jack I should not have taken under any consideration whatever."

"Why, captain, I thought Jack and Odell was two of your best men. What have they done to get out of your good books?"

"Well, Jack, you see Cape Town [Jack] cut out his flukes about [a] fortnight ago and we drove him away from the shanty."

"And how about Odell, captain?"

"Oh, he is another one; we drove him from our shanty and he

has been living in the men's shanty for about three months."

"Captain Fuller, did I not tell you that them two suckers was a damn sight worse than I am and that you would find it out some day?"

"Yes, Melrose, I remember you said that they was two rascals and I have found it out so too."

"Well, captain, I am glad that I went over there when you had that row with Cape Town Jack, but I have it in for Cape Town and some day I will pay him off and don't you forget it, captain."

"Jack, when we get to the cape them fellows will want me to let them have some money, you see if they don't. But they will find want their master."

"I should think, captain, Odell ought to be ashamed of himself and he ought to [have] tried to get along with you folks, he living, you may say, next door to you and old man Chipman."

"There is no shame in him, Jack; you see if he don't fetch up in prison one of these days."

"Yes, captain, unless he gets his head knocked off. He thinks there is no one like Cape Town Jack."

"Melrose, do you and Fink want to go over there and live until the *Francis Allyn* is ready to leave for the Cloudy Islands?"

"Not by a damn sight, captain; I will stop where I am. We don't want to have nothing to say or do with them fellows."

"You are right, Melrose, stop where you are and the day before we leave I will come after you two."

"Captain, who be you going to put on the Cloudys?"

"I am going to put Mr. Fuller, Manice, cook, steward, Manuel the Portuguese, and you two; that will make up a boat's crew and one over. And Captain Glass is going to put two boat's crews there."

After a little more talk I left them and went on board of the *Francis Allyn*. When I got on board, Captain Glass wanted to know how I made out. I told him that them two was all right and they would make up a good boat's crew and one over to put on the Cloudy Islands. "And now, Captain Glass, whenever you are ready we will draw up the agreement."

"I am ready now," said Captain Glass, "We might as well do it now as anytime." And this is a true copy of the Agreement we drew up and signed:

Nortons Harbor
Island of Desolation

September 13, 1882

This is to certify that I Robert H. Glass do agree with Joseph J. Fuller to give one third of this seasons catch of sealskins the officers and crew reserve their respective lays remainder to Joseph J. Fuller

September 13 1882   John Thompson   Robert H. Glass
                    Oscar Myers   Joseph J. Fuller

Sometime the first of October, we was all ready to leave Norton's Harbor for the Cloudy Islands. I went down to Atlas Island and got Melrose and Fink and fetched them on board of the *Allyn*, where I had [an] agreement drawed up stating what lay or share each one was to get out of the net proceeds of one-third of the sealskins that we should take this season from the Cloudy Islands or any other place around Kerguelen Island, each one signing his name to it. After the men had signed the agreement I said to Captain Glass, "There, Glass, we are ready for work." [At the] same time Captain Glass said to me, "Captain Fuller, how do you think it best to work?"

"You see, Glass, it's not for me [to tell] you [how you] should work; you are master of [your] own vessel and I should not like to have your officers think that I am interfering with your voyage. I am willing to do whatever you say."

"What would [you do] if you had the management of the voyage?" said Glass.

"Well, captain, I will tell you what I should do if I was in your place. I would land one boat's crew with mine, that will make fourteen men, on the Cloudy Islands, and that will be men enough to take all the seal they will get, I am thinking. Then I would take the vessel and the remainder of the crew and go to Thunder Harbor and fill her with oil."

"Yes, Captain Fuller, that is all right enough but [what] are we going to do with the oil if we should get two or three thousand skins?"

"Why, Glass, you can easily put the oil on shore to make room for the skins."

"Captain Fuller, I lost one season's work at Cape Horn by trying to do two things at once and I think we had better put three boat's crews there and wait and see what the prospects will be before we get oil."

"Do what you think is for the best. It will be all the same to me. But you can fill the *Allyn* with oil before the season fairly commences for seal and then if there is not as many seal as we expect you will have your oil to fall back on."

"I think I shall hold on a while and see what the prospects is, but we will pick up what elephant we can in the meanwhile."

"Captain Glass, as you are all ready to leave I think you had better get off [the beach] my two boats and I think you had better let your officers go in with mine for I think them rascals of mine will try and stop them from taking the boats."

So Glass told Mr. Usher what I wanted.

He said, "All right, Captain Glass and Captain Fuller, they will find that we are not to be fooled. I should like no better fun than for them to try and stop me and if some of them don't get sore heads my name is not Usher."

When the men had all got in the boats I told them that I thought I would go too, so I got into the boat and we [went] ashore. We landed alongside of the rocks and walked up to where the boats was and began to take off the fastening. All the men came out of the shanty and came to where we was and Odell and Carroll said, "Be you going to take the boats, Captain Fuller?"

"Yes," I said. "Don't you see what the men are doing?"

"If you are going to take the boats, captain, what are we going to do for a boat?"

"I don't care what you are going to do for a boat, Odell."

"Captain Fuller, the boats belong to us as much as they do to you."

"If you think so Carroll, why don't you stop us from taking them?"

Melrose spoke up and said, "Let them try it on, Captain Fuller; they think themselves damn smart, but you have got the deadwood on them suckers this time. If I was you I would give that Odell a dig [on the] side of the head."

"You think yourself smart, Jack Melrose, because you have the captain at your back," said Odell.

"Yes, I am smart enough for any one of you suckers, and, if any of you fellows want to try me, take off your jackets and come. I am going to back the old man up about these boats and don't you forget it."

Not one of them had the pluck to take Melrose up, but Carroll said, "Why cannot you let us have one of the boats, captain? How be we going to get blubber to burn, captain?"

"I don't care how you get blubber to burn. Captain Glass made you a good offer and you rejected his offer. Now you can get along the best you can. I want you to know that some of my money bought these boats and now you stop my men from taking the boats at your peril."

"We helped to save the boats," said Odell, "And they ought to belong to [the] largest party. You have no more right to the boats than we have."

"You think so, Odell? Well, you thought the same way about the sealskins when you and that Cape Town Negro took them out of the cask and took them inland and buried them about a mile from where they was found. Jack Melrose knocked your little game in the head for you and now you want me to leave you the boats so that you and the rest of you can go to the Cloudy Islands and interfere with our season's work. Don't think that I don't know what you are up to, Odell."

"That is just what they are up to, Captain Fuller," said Melrose, "for they wanted me and Fink to join them when they made old man Chipman come over that time to Atlas Island."

I told Mr. Usher to shove the boats down into the water and I

said, "Odell and the rest, now you try and stop them and we will
see who is the best man." But they thought better of it. All they
did was to stand and the men run them into the water. After the
boats was in the water I said to the men, "You have had a good
offer and you have seen fit to reject it. Now you can stop here on
the beach until Captain Glass is ready to take you to the Cape.
There is the provisions that is left from the *Pilot's Bride*. Use it but
don't take nothing from the *Francis Allyn*'s stuff, for if you do I
will make you pay for it for you all have money coming to you.
For the next two or three months I shall have nothing to do with
you, as I am going on board of the *Allyn*. You can live in peace or
you can fight and quarrel as much as you please and do what ever
Cape Town Jack tells you to do." I then went to [the] boat and we
all returned to the schooner, taking the two boats with [us].

The next morning Captain Glass got underway from Norton's
Harbor and worked up to Fuller's Harbor. He [had] the boats in-
shore all the way up, killing elephant and leopard and getting the
blubber on board. We laid in Fuller's Harbor two or three days
until we got all the elephant and leopard there was to get and
while lying there we got [a] large number of young albatross from
Howes Foreland. From Fuller's Harbor we went to Breakwater
Bay. After getting what blubber there was, we boiled out the
blubber and stowed it down in the hold so as to have [a] clear
deck. We then got underway and worked the islands of Cumberland Bay Head. We got a few seal there. That afternoon we came
to anchor in Christmas Harbor, where we had to wait for the
wind and weather to come around right to get to the Cloudy Islands. While living there we got two seal.

About the middle of October the wind came out light from the
southwest at dark and at four o'clock in the next morning Captain
Glass got underway and worked out of the harbor. The wind
being light, it took us until daylight to do so. Some of [the] time it
would drop off [to] a calm and then a heavy puff would come that
would lay the *Allyn* down to her planksheers[37] and before she
could gather headway it would drop off to calm again. At last we
drifted out clear of the heads and got the wind steady from the
westward and strong. We worked up under the mainland, making

one long and one short leg until we got abreast of the Cloudy Islands. We arrived at the Big Cloudy about noon, where we was going to land the gang at [a] place we called Tony's Harbor. The gang was not to go on the small island where we got our seal in the *Pilot's Bride* until December fifteenth, but they was to kill and get all the seal wherever [they] could find them on the other islands and rocks.

Soon after [we] got there, the wind chopped around to the northeast, so we took all five boats. In one hour we had them landed with all their things, salt, and provisions. We landed Mr. Fuller with six men; Mr. John Glass and Mr. Enos, third mate, with ten men, making nineteen men all told. As soon as we landed them we ran back to Christmas Harbor. From there we went back to Fuller's Harbor and waited for an opportunity to work Swain's Island and Terror Reef. We got [a] few seal on each place.

When we came back to Little Harbor, that night I asked Captain Glass again about getting his vessel full of oil. "Oh, wait a while longer and let us see what seal there is going to be first," said Captain Glass. But I did not give him no rest until he said that the last of November he would go to the Cloudy Islands and see what they had down, and if it did not look favorable he would go in for oil.

"Captain Glass, you can take five hundred and fifty barrels of oil and then have plenty of room for your water and skins. The oil will be worth ten thousand dollars and that will be all clear gain to you and your crew for you know that my gang have no share in the oil that you take; but that will make no difference to us, we will work and help you to get the oil all the same."

When the last of November came around we went to the Cloudys and Mr. Usher went into Tony's Harbor and got Mr. John Glass and his boat's crew off from shore. And when he came on board Captain Glass asked him what they had done. He said that they had taken about three hundred and fifty sealskins. "Mr. Glass, what do you think of the prospect for seal on the small island?" I said.

"Captain Fuller, Mr. Fuller and myself went over there a day or two ago. There was about four hundred whigs up but not a great number of female seal. If there was going to be any great number of seal they ought to be coming up fat about now. But I think if we get the same number of seal that you got the first season in [the] *Pilot's Bride* we will be doing well."

We landed some more provisions and left word with Mr. Fuller and Mr. Enos that we was going to fill the schooner [with] oil. The wind being light from northeast, we kept away for Thunder Harbor and we got clear of the Cloudy Islands. The wind dropped off almost to calm and we was out all night. That night about midnight I heard the watch sing out, "Breakers!" and in a minute or two I heard Captain Glass say, "Call Captain Fuller." By that time I was dressed and I went on deck. As soon as I got on deck Glass said, "What breakers is them on the lee bow, Captain Fuller?"

"Why, Glass, them are the breakers of African Head and we are drifting down onto them. You had better get the boats down and get them ahead and help her with this light wind to work by, but we are some ways from the reef."

They got two boats down and began to tow, but they did not seem to make much head way. So I told Captain Glass he had better let Mr. Chipman take the other boat and help tow.

"All right, Captain Fuller, I wish he would," said Glass. As soon as the third boat got hold and began to tow we could see that the schooner was working off from the reef fast and in half an hour the schooner was out of all danger and [a] light breeze sprang up. I told Captain Glass he had better take up his boats for he was out of all danger and let her come to the wind and lay to for the remainder of the night, for we was up far enough. So he called the boats alongside and hoisted them up and waited for daylight. After taking the boats up Mr. Usher said to me, "Captain, them are nasty breakers."

"Yes, Usher, and I should not like to get on them; I have had enough of shipwreck life to last me the remainder of my days." But there was no danger of us getting on them, for we could have gone in between them and the mainland; there is plenty of room.

The next morning at daylight it was fine, with light wind from northeast. We kept away for Thunder Harbor, which was about

twenty-five miles away, and as we drawed into the land the wind
keep freshing. At noon we came to anchor in Thunder Harbor
and that same day we moored the schooner, putting the big anchor to the northeast and paid out one hundred and thirty fathoms
of chain on it. Then we let go the small anchor to southwest and
then we hove back in the big chain until we had hove in seventy
fathoms, that gave sixty fathoms on the big anchor and forty-five
fathoms of chain on the small anchor. After getting the schooner
moored, all hands went on shore killing and skinning elephant.
When we knocked off work for the night, we fetched the blubber
off that we had skinned in rafts and let it lie in the water alongside
of the schooner so as to soak out the blood from the blubber before
mincing and boiling it.

After killing that day and the next, I told Captain Glass that I
thought we could keep him boiling and mincing with two boats'
crews if he wanted to commence boiling. But he thought I had
better take three boats' crews, as that would leave him ten men.
The next morning being fine I told Mr. Usher that we had better
take the three boats up to Shoe Foot killing, so that when we had
done killing we could fetch down a raft of blubber the same day.
When we got up to Shoe Foot, we went in back and went to killing. We kept to work so busy that we did not mind that the
weather had changed when about two o'clock Mr. Glass went
down to the boats for something and when he came back he said to
me, "Captain Fuller, you had better have a look at the weather.
It's blowing strong outside and there is a heavy sea coming in the
bay and breaking almost across the passage."

I went up on a little hill and had [a] look off into the bay. I
only took one look; that was enough for me. Outside the bay the
water was feather white and a heavy sea was breaking on the
beach. I hurried back and told the officers that we had better bury
the blubber as quickly as we could and get on board of the
schooner, for it was going to blow a heavy gale of wind. So at it
we went, all hands working with [a] will. In [a] little while we had
the blubber all buried and we all went to [the] boats. The sea had
got worse and it was breaking right across the passageway. First
one officer would look at it for a little while and shake his head and

say, "You don't get me out there in a boat," and then another, "Nor me either! I would rather stop on the beach all night than try to get on board in that sea." I let them talk until Usher said, "What do you think of it, captain?"

"Well, Usher, it will be hard work getting on board with the sea breaking across the passage the same as it [is doing] now. If they had [a] boat on board of the schooner, we might get on board but I think it doubtful [even] then for I think it's blowing heavy where she is laying and they could not pull one on shore if they had one."

"Well, what is the best to do, captain?" said Usher.

"I will tell you what I think. We had better carry the lightest boat across [the] land to where the schooner lays."

"How far is it across?" said Usher.

About [a] mile and [a] half, I should think and it would be better than stopping all night on shore and perhaps we can not get off tomorrow. What do you say about trying?"

"If you say the word, captain, we will make the wool fly."

"I say we had better. There is twenty of us; that will be about seventeen men to the boat and three to carry the oars and things."

"All right, captain," said all of the officers, "But what about the other boats?"

"Haul them well up on the beach and lash them together and fill them with blubber and they will be all right."

"But, Captain Fuller, won't the birds eat the blubber out of the boats?" said Usher.

"No, Mr. Usher, they will not trouble the blubber as long as the blubber is in the boats."

We hauled the boats well up from the water and lashed them together and filled them with blubber and then we divided the men half on each side of the boat and three of us taking the oars. The sixteen men picked the boat up and walked away with [it] for about five hundred yards; then they would set her down and take a rest for five minutes. Then off again they would [go] with her. We done very well as long as we had hard walking. But when the men came to where the ground was soft and marshy then came the tug of war. The men would go to their knees in mud and water. It

made it hard work and the men had to stop more often and now
and then one would go to his waist in a hole; then he would have
to stop and swear.

But as all things must have an end, so did this. We came to the
beach ahead of the schooner and it was blowing big guns, coming
in heavy puffs. We got the boat into water and all hands got into
her, keeping her head to the wind. When a heavy puff would
come we would pull [her] ahead, holding her until it passed by;
then we would let her drift down again, holding her when a heavy
puff would come. At last we got alongside and safe on board. The
next day it blowed heavy until sunset. From that time we had
very good weather and by the twentieth of December we had all
the oil wanted, about five hundred and fifty barrels.

On the twenty-third we left Thunder Harbor and ran down to
the Cloudy Islands, but we found it too rough to land so we kept
away and ran down to Fuller's Harbor. When we came to anchor
some of the men went after penguins' eggs and when they came on
board they said they had seen men tracks. I told Glass I thought
my men had made a boat and I thought they had tried to get to the
Cloudy Islands. They had got as far as Fuller's Harbor then they
had to give it up as a bad job.

The next day [we] got under way and ran down to Norton's
Harbor. When we came to anchor some of my men wanted to
come on board. When I saw them coming I told Glass I wished he
would allow them on board. They had built themselves a small
boat covered with canvas. When they came alongside, Glass told
them that he had the smallpox and they could not come on board.
So they went on shore again.

The next day I went on shore and I had a look at their boat
that they had built. As I was looking her over, Jim Magill come
along to where [I] was. I said to him, "Well, Jim, you fellows have
got a boat, I see, but you did not go on her to Fuller's Harbor, did
you?"

"No, sir, the fellows built a larger one and they went in that
one. They was trying to get to the Cloudy Islands in her. When
they got to Fuller's Harbor, they gave it up as a bad job."

"Did you [go in] her, Jim?"

"No, sir, I wasn't damn fool enough for that. Captain, when they was coming back, they came near getting lost."

"How was that, Jim?"

"You see, sir, when they was coming back, when they got in the middle of the north arm the wind struck down a heavy gale and they could not get back to where they started from. So they had to keep right before the wind, and they just fetched the Jug Keys[38] and they had to lay there under the lee all night until it moderated; and you cannot get one of them to go in her again. They have had enough of her, I can tell you. When they came back they was a scared lot of men."

"Where is the boat now, Jim?"

"She is around in King Harbor, hauled up on the beach."

"Who was boss of her, Jim?"

"Why, captain, you might know without asking, Cape Town Jack and Odell."

"I thought as much, Jim. It's a great pity that them two had not been lost in her; but I should not like to have seen anything happen to the other men."

That same day Captain Glass got some more provisions on board and we left for the Cloudy Islands again. We worked Swain's Island and Terror Reef, but we only got fifteen sealskin on both places. Then we came to anchor in Breakwater Bay, where we laid until the wind came around light southwest. We got underway and worked out to the Cloudy Islands and took off the men and sealskins. They had take 1,175 skins, making with what we had on board all told 1,255 sealskins for the season's catch. From there we went back to Norton's Harbor and got ready the vessel for the Cape of Good Hope, taking in water and provisions for the passage. On the fourteenth of January we all of the crew of the *Pilot's Bride* went on board of *Francis Allyn,* she being all ready for sailing. When the crew came on board I told them that [they] must help work the schooner. On the sixteenth of January 1883 we left Kerguelen Island, making fifteen months and twenty-three days since leaving home.

We had a very light passage to the Cape of Good Hope, where we arrived February the thirteenth, everybody in good health.

That same day after getting in, Captain Glass sent a telegram stating his success and the finding of the crew and the loss of the *Pilot's Bride* and four men. The next day Glass had a telegram back from Williams wanting to know who was lost. So I answered it back saying, "Gray and Cole and two Portuguese," signed "Fuller."

As soon as I got into Cape Town I turned the men over to the U.S. Consul and got them off my hands, he giving them board and clothing, [the] which last they was in much need of. But for myself he would do nothing. He said I was an owner's man and government did nothing for masters. But I found some good friends in Cape Town, more specially the firm of James Seawright & Co. They done everything for me that men could do for another man. They even went to the agents of the steamer *Athenian* and interceded for my passage in her to England for first-class passage, I paying second-class passage, and giving me a letter to the agents of the North German Lloyd Steamboat Company. Then they wanted to know if there was anything else they could do for me. I told them no, they had done all they could and I thanked them for what they had done and I never could repay them for their kindness. They said, "Captain, we don't [do] this for pay; we do it for the good will we have for you and don't thank us either." I shall always remember them with gratitude.

A few days after we arrived at Cape Town about all of the men that would not take Captain Glass's offer came to me and wanted to know if I would let them have some money. I asked them if they remembered what I told them at the Kerguelen Islands. They said yes, "But, captain, let bygones be bygones and forget and forgive."

I told them that I had [not] forgotten them or their conduct either, and I should not let them have no money. They even came to me when I was going on board of the steamer, but they got no for [an] answer. But all them that took Captain Glass's offer, I let them have money. Just five days after arriving in Cape Town, Mr. Chipman and myself was on [our] way to England. I arrived about the twentieth of March. On the passage nothing could exceed the kindness of the captain, officers, and passengers to me; I shall ever hold them in grateful memory.

I stopped at Southampton a week, waiting for [a] steamer. As soon as I arrived in Southampton, I presented my letter to the agents. They charged me for second-class passage and Mr. Chipman [for] steerage passage, and they gave me a letter to the captain of the steamer *Warren.* As soon as the steamer arrived, I went on board and presented the letter to the captain. After reading the letter he said, "Captain, take your things into the second cabin for now and I will see you later on. I am very busy just now." That evening he sent the head steward to me, asking me to come into his cabin. I went into the first saloon, where he was, and he said, "Captain, you let the steward have your traps removed into the next room to mine and tell your mate to take your room in the second cabin." Then he wanted to know where I lost my vessel and how. After telling him where and how I lost her, I said, "Captain, I thank you for your kindness to me and my mate."

He said, "Don't thank me, captain, I am only too pleased to think that I can help you, for you know we do not know when we shall want the same favor shown to us. That is in our profession and if there is anything that I can do for you, let me know."

I said, "Captain, you are very kind and I thank you for your kind offer all the same." After leaving the captain, the steward showed me my room where I found all of my traps and the steward said to me, "Captain, you will find your mate in your room in the second saloon." So I went in the second saloon and I found Mr. Chipman. I asked him if he was satisfied.

"Well, captain, I ought to be after stopping one day in the steerage."

"You ought to thank the captain, Chipman, for his kindness."

"Oh, you thank him for me, captain."

"That is done already, Chipman," I said.

"Captain, I will tell you what I will do. I will take a drink on it for luck and that will do just as well as thanking the captain."

"All right, Mr. Chipman," I said, "we will go to the smoking room." When we got into the smoking room, the room was full. Someway the passengers had found out that there was two shipwrecked men on board, captain and mate. They knew us as soon [as] we got into the room and they all flocked around us, introducing themselves. After a while they wanted to [know] what my mate and myself would have to drink, "And, captain, we want you to understand that you don't pay for nothing as long as you are on board of this steamer. Now what is it to be?" I told them what we would have; but I thought to myself that they would not have many drinks to pay for me but as for Chipman, he would take all they might give him and he would ask them to take a drink with him once in [a] while. From that time, whenever they did see me on deck, someone would ask [me] into the second saloon and the first was just the same. The ladies was the same, but instead of drink they would stuff me with fruit and cake and they would not take no for an answer.

From Southampton we was eight days to New York, where we arrived at night and we stopped on board all night. The next morning we started for New London. When I got to New Haven I sent a telegram to Mr. Crandell,[39] telling him of my arrival and to meet me at the New London railway station. When I arrived at New London I could see nothing of Mr. Crandell; so I got out of the cars on to the platform, but I could see nothing of him. Seeing no one, I went down on Water Street where the offices was, but I found the offices shut up. As I was returning to the station I met Mr. Crandell coming to the offices. He was mighty glad to see me. Someway we had missed each other at the station. He said, "I knew you would go to the offices, captain, if you did not see me at the station, but I was afraid to leave the station until the cars [had] gone. You have missed this train. You might as well come home with me and take tea until [the] midnight train comes along."

As we was going to his house, I met a great number of friends. They was glad to see me and they all congratulated me on my safe return. After tea I gave Mr. Crandell and his wife the particulars of the loss of the *Pilot's Bride*, and how we lived, and the trouble I had with the men; and I gave him the bills and other papers. I told Mr. Crandell that I should like to go over to East New London to Captain Glass's house, as I had three hours to wait, if he would tell me where Mrs. Glass lived. So he went over there with me.

When we came to the house and he was bidding me good-bye,

I said, "Mr. Crandell, when the *Allyn* comes in if you will let me know I will come down."

"I wish you would," he said; "By that time I think Mr. Williams will be home from Europe and he will want to see you." So I wished him good night and went over to Captain Glass's house. She was very much pleased to see me and to hear from Captain Glass. I stopped there until about eleven o'clock; then I went to the railway station, where I had [to] wait until midnight for the train. As soon as the train came in I got on board and found an empty seat and I made myself comfortable and went to sleep, only waking up when I arrived in Boston. I went across the city as quickly as I could, but I found after getting to the Boston and Main Depot that I would have to wait three hours before a train would [go] direct to Danvers, but only half an hour for one going by the way of Salem; so I took that train and I arrived in Danvers about eight o'clock.

When I got out of the cars I found a large number of friends and acquaintances, and it was hard work to get away from them until the only hackman that was there got me by the arm and said, "Come, captain, let me get your traps, for there is a little woman on the hill waiting for you for she expected you last night."

"Why how is that? I sent no telegram of my arrival."

"Well your owners did yesterday, saying that you had arrived in New York." He soon had my traps in the hack and I got in. As we went through the streets past the shops, the proprietors was standing in their doors or sidewalks waving their hands and hats, but the hackman would not stop for none and in about ten minutes he put down at my gate. As we stopped at the gate, my wife was just opening the side door that opens on the porch. She had a broom in her hand. As soon as she saw who it was the broom went out of her hand in [a] hurry and, kind reader, you can imagine what followed — first [a] little crying and then laughing from wife and little ones.

That evening some members of the G.A.R. Post Ninety of which I was a member called on me. When they was leaving they said, "Captain, next week's our regular meeting of the post and we

expect you will be there and give us a full account of your shipwreck experience." I told them that I was no talker and that I had
rather hear someone else talk than myself.

"Well, captain, the post [does not] expect a Webster of you,
but you can talk well enough to give the post an account of the
shipwreck and what you had done for provisions while you was
on the island." I saw it was of no use to argue with them; they
would not take no for an answer. At last I gave in and told them
that I would be there. Soon after they had done, Mr. Damon,[40]
one of our neighbors that lived across the street, also a member of
the G.A.R., came in. After congratulating me on my safe return,
[he] said "Captain, you don't [know] what a stir there was in the
Post when your wife got your telegram saying you was in the
Cape of Good Hope and was well. You see it was this way. That
evening before going to the post, I happened to see your wife out
by the gate and she said that she had heard from the owners by
telegram that you was found and expected you home in about a
month's time. After hearing all she could tell me, I went to the
post meeting. I happened to be late that evening so I said nothing
until the post had got through with their regular business. I got up
and said that I was most happy to tell the post that Mrs. Fuller
had had a telegram from the owners of the *Pilot's Bride* saying that
[the ship] was lost and that Captain Fuller was in Cape Town and
was expected home in about a month's time. And Fuller, before I
could get through speaking every member was on his feet and
such a cheer went up for about two minutes that it did one's heart
good to hear. And then the post had [a] letter of congratualtion
sent to your wife. And now, captain, the post has made arrangements for a big campfire for you, to have you at our next meeting
tell them of your shipwreck and you must not disappoint the
boys."

I told him that I would see when the time came. When he had
gone I told my wife I thought the post would have to wait, but she
thought I ought not to disappoint the post for they all had been
very kind to her, sending and wanting to know if she was in need
of any assistance, and if she was, all she had to do was to let them
know. And she then gave me the letter of congratualtions to read

that the post had sent her. After I had read it, she made me promise that I would go and give the post an account of my shipwreck life. I told her that I felt so happy to be at home again that I could not say no to nothing she might ask of me. When the evening came for the meeting of G.A.R. Post, the post sent three delegates to my house to escort me to the hall; and when we came to the hall and was ushered in and presented to the commander of the post, I saw that the hall was packed full and at [a] sign from the commander every comrade rose to his feet and at another sign they gave me three times three. And I thought to myself, "Well, Joseph, you have put your foot in it this time," and if the commander had given me time I think I should have been stage struck.

After receiving a congratulatory address from the commander of the post, I was asked if I would give them some of my shipwreck experience. I was invited to the platform, where the commander introduced me to the comrades for there was some comrades [there] from other posts. After being introduced I told the comrades that they must not expect too much of me for I was no talker but I would do the best I could and that I should have to begin at the beginning of the voyage so that they could understand the whole thing. I told them about leaving New London, and of the places that I stopped at before I arrived at Kerguelen Island, and of my taking oil, and of my finding the seal and the number of seal that I took off the island, and how I had to go to the Cape of Good Hope after salt, and of my going back to finish up the seal, and of my finding the bark *Trinity*'s provisions and other stuff in Pot Harbor sometime in July. By that [time] we thought that something had happened to her, and we had made it up that in October we would go over there and see what had happened to her after landing [a] boat's crew on the Cloudy Islands if there was no vessel out from home looking for her. And of my losing the boat and four men, and by looking for the bark *Trinity* I lost my own vessel, and of my journey from Pot Harbor to Royal Sound to leave a letter, and of my troubles with the men, and the day of days when the *Francis Allyn* arrived, and from that time until I got home.

After I got through with the narration the comrades gave me a vote of thanks. The next day my wife was down the street; when she came home she said, "I met our old family doctor and he said to me, 'Why, Mrs. Fuller, I did not know that the captain had so much gab.' I said, 'Why, doctor?' 'Why, Mrs. Fuller, I was at the post last evening and he did first rate.'"

I told her that I was jolly glad when I got through with the whole thing.

A fortnight after I got home I had [a] letter from Mr. Crandell saying the *Allyn* had arrived home and also Mr. Williams from Europe and he would like to see me. After receiving his letter, in [a] day or two I went to New London. Mr. Williams was very glad to see me. After some talk he wanted to know when I was going back home. I told him that I thought of returning that same day, "But if you want me to I will stop over."

"No, no, captain, you be with your family all you can, but I should like to know what [you] think of doing."

I told him I suppose I should have to go to sea again, as I had to do something for [a] living.

"Well, captain, it will devolve on us to get another vessel for you."

I said, "Well, Mr. Williams, I don't know if you want to buy another vessel or not but what be you going to do with the *Francis Allyn* as Captain Glass has come home sick?"

"Will she do and will you take her?"

"I think she will do. I should like to have [a] look on some of the other islands around Kerguelen Island if I can and I can have Captain Glass's part of her, if he is going [to] sell."

"You can have Captain Glass's part, captain, if you will take her." I told Mr. Williams that was all right and he said, "Now, captain, when do you want to sail?"

I told him that I thought that if he thought of sending her he ought to get away by the last of July, "But, Mr. Williams, I should like to have you say nothing about sending a vessel down to the Kerguelen Islands for the Lawrences are going to send the *Colgate*. Her captain came to me and he wanted to know if I was going out this spring and I told him that I did not think that I should this spring."

"But, captain, will July be time enough to get there?"

"Yes, plenty of time; you see, Mr. Williams, the captain of the *Colgate* thinks that there will be no one there but himself, and he will not trouble the Cloudy Islands until he gets the *Colgate* full of oil, and by that time I will be down there and have all of the seal. I will, after the first killing, leave [a] boat's crew there and take the remainder of the crew and go and fill the schooner with oil."

"Well, captain, I think that is well planned and we will not say nothing about going until the *Colgate* sails." At the same time Mr. Williams wanted to know how much the provisions was worth that I took from Lawrence & Co. I told him I thought two or three hundred dollars but they was worth to me about six hundred dollars: that would be sea price.

He said, "Very good, captain, [I] will make them that offer but they are hard people to deal with."

Sometime in July I was in New London getting ready to sail. Mr. Williams asked me if I had my agreement that I had with Captain Glass, for he said that Captain Glass had lost his. I told him yes, I had mine. He said, "Will you let me see it, captain?"

I said yes, and when I came to the offices the next day I took it along with me and I gave it into his hands. After reading he said, "Well, captain, this agreement is good for nothing. When you made this agreement you ought to have written it this way: 'Captain Robert H. Glass of schooner *Francis Allyn*' instead of 'Robert H. Glass,' and yours should be 'Captain Joseph J. Fuller, ex-captain of the schooner *Pilot's Bride*.' And another thing, Captain Glass says he thinks that he ought not [to] give you one-third of the sealskins. And captain, I think he is right but I am willing to give you $1,000 for your help."

"What, Mr. Williams, does Captain Glass say? That he don't think I ought to have one-third of the skins? After what I have done for him in helping in getting his voyage! I know this much, Mr. Williams, that I did more to get his voyage than [he] himself or any mate he had on board of the *Allyn* and now he says that he ought not to give me one-third of the skins after he making the offer himself. I always knew that he was mean, but this is the meanest thing yet if he is my uncle. And Mr. Williams when I made the agreements with Captain Glass I knew that I was no lawyer but I thought that the owners of the two schooners would overlook all legal points and give me what the agreement says without any trouble."

"I think, captain, $1,000 is [a] fair offer," said Mr. Williams.

"You may think so, Mr. Williams, but I do not and if you only knew the trouble I had with the men to keep them from going to the island and destroying the seal you would think so too, and if Captain Glass had not said that he would give me one-third of the skins I should have taken the boats and gone to the Cloudy Islands and Glass would not have taken one hundred skins for his season worth in all."

"Do you know, captain, that Captain Glass would have been justified in shooting you or your men for going on the island and destroying the seal?"

"That might be so, Mr. Williams, but I doubt it very much. The seal belonged to me as much as they did to Captain Glass or anyone else until they was caught and they would not shoot but once before they would have been cleaned out."

"Well, captain, I think Captain Glass ought to have some recompense for going after you. All the government allowed him was fourteen dollars a man."

"I think, Mr. Williams, Glass was well paid for his trouble, his voyage amounting [to] something over thirty thousand dollars. His share is some six thousand dollars for nine months' time: I think that is good pay."

"Well, captain, I do not know what you would have done if Captain Glass had not gone after you, for I do not think the government would have sent a vessel after you for they had enough when they went after the *Trinity*'s crew at Heard Island, and, captain, I do not think that I should fit out a vessel for that amount."

"You say, Mr. Williams, that you would not fit out a vessel for thirty thousand dollars? How much did her voyage to Cape Horn fetch? Only something like thirty-six thousand dollars and you owners thought she had done well and she was gone eighteen months and this time she was gone about nine months and her voyage amounted to about the same."

"Well, captain," said Mr. Williams, "I don't want you to think that I am doing this for my own benefit for I do not own one dollar in the *Francis Allyn*, but, if you think you can get more than what I offer you, you can take it into court and see if the law will allow you more."[41]

I said, "No, Mr. Williams, I shall not take it into no court; but if I had known when I was in the Cape I should have made Glass given me my share there, for he wanted to know if I wanted to divide the skins there. But I told him that he had better let it be until we got home. By doing [so] I am the loser, but you can tell Captain Glass I will make him [a] present of the skins. He is rich and I am poor; even [if] he is my mother's brother, he is worth about fifty thousand dollars and I am worth about as many cents. But, Mr. Williams, you will never catch me that way again."

But I will leave it with the reader to pass judgment on the owners and Captain Glass's actions in the case. At the same time Mr. Williams told me that Mr. Lawrence had refused the offer of $600 for the provisions and they wanted $1,300 but [he had] let them know that [he would] not pay them no more than $600. We heard nothing more from them until the day before I was to sail. That day, along in the afternoon, I was on the wharf talking to Mr. Williams about the voyage when Lawrence came on to the wharf. After bidding Mr. Williams good afternoon [he] said, "Mr. Williams, I have come down to see you about the provisions that your people have taken belonging to me at Kerguelen Island."

"Very good, Mr. Lawrence, come into the offices," and as he was going in he said to me, "Captain, you come in too, will you not, as you know all about the provisions?" So I followed them in in company with Mr. Crandell. When we got into the office Mr. Lawrence said, "What are you going to do about the provisions that your people in the *Pilot's Bride* took belonging to me?"

"Why, Mr. Lawrence," said Mr. Williams, "I made you the offer of $600 for what they took."

"I will not take it," said Mr. Lawrence, "And I want $1,300; I want sea price."

"Well, Mr. Lawrence, I will give you $600 and not [a] cent more and if you think you can get more you can try it in the courts

and I think $600 is sea price." Mr. Williams turned to me and
said, "What was it you took, Captain Fuller?"

Before I could tell him Mr. Lawrence said, "I don't want to
know what Captain Fuller, a damn thief, knows about the provisions."

That was enough for me. I went for him and if Mr. Williams
had not interfered there would have been a small circus in that
room. Mr. Williams told Mr. Lawrence that if he could not behave himself as a gentleman there was the door, for he asked Captain Fuller to come in so as to give information about the provisions. "You can do your business and go."

Mr. Lawrence got up from his seat and said, "Mr. Williams,
you say that you will not give me my price for the provisions."

"No," said Mr. Williams, "I [will] only give you the $600 and
no more."

Mr. Lawrence [headed] for the door. Before he got there he
turned around and said, "Mr. Williams, if you don't pay the
$1,300 your vessel shall not go to sea tomorrow."

Mr. Williams said, "She will go to sea about nine o'clock and
you can stop her if you can." With that Mr. Lawrence left without
saying another word.

After Mr. Lawrence had left Mr. Williams said, "Captain, I
am glad you did not hit him and kept your temper under control.
He wanted you to hit him or do something so that he could get
some hold of you to stop you at home as long as he could so as to
give his vessel the *Colgate* [a] chance to work the Cloudy Islands.
He will try and stop you if he can do so, but you come to the offices about half past eight o'clock tomorrow morning and I will
have things all right."

The next morning I was at the offices at the appointed time.
There was no one there but Mr. Crandell and he told me that
Lawrence had got out a warrant for my arrest last night and he
wanted the sheriff to arrest me last evening but he could not get no
one to serve the warrant. In a few minutes about half a dozen
friends of Mr. Williams, the money men of New London, and the
sheriff [arrived]. As soon as the sheriff came in, he walked up to
me and asked me if my name was Joseph J. Fuller. I said, "That is

my name." He took out his warrant and said, "Joseph J. Fuller, I arrest you in the name of the Commonwealth of the state of Connecticut." And Mr. Bond,[42] one of the gentlemen said, "Let me see your warrant for the arrest of Captain Fuller." The sheriff produced the warrant and handed [it] to him. After reading it he asked the sheriff if he had the bond papers with him. The sheriff said he had them with him.

In a few minutes after the arrest Mr. Williams came. Seeing the sheriff standing in the offices he said, "Captain, they have got you have they? Well, we will soon have that fixed all right," and he said, "Sheriff, let [me] see the papers." After reading them he said, "I see, captain, they have attached your house for $2,000." Then Mr. Henry Bond said, "I will be one of Captain Fuller's bondsmen, Mr. Williams," and the rest of the gentlemen offered to do the same. But Mr. Williams said to the gentlemen, "Gentlemen, I thank you on behalf of Captain Fuller and myself for your kind offers, but I shall be Captain Fuller's bondsman myself." After he had signed the papers and the sheriff had left, Mr. Williams said, "There, captain, you go to sea and don't let this worry you one bit for it shall not cost you one cent whichever way the case may turn."

I thanked Mr. Bond and the other gentlemen for their kindness and they bid me good-bye and God's speed. As this is the last of the *Pilot's Bride* voyage I will finish it up at once. About the time I was expected home on that voyage, Mr. Williams notified Mr. Lawrence that he wished to have the case fetched to a trial or settled, so Mr. Lawrence, rather than take the case into court, took the $600, so that it all ended in smoke. And, kind reader, that is the last of the *Pilot's Bride*. [If] this should ever appear before the public I hope they may take as much pleasure in reading of the account of the shipwreck as I have in putting it on paper in my leisure moments and the only thing that I regret is that I am so poor [a] storyteller.

Appendices, Notes, and Selected Bibliography

APPENDIX 1 VOYAGES OF JOSEPH J. FULLER

SOURCE: Alexander Starbuck, *History of the American Whale Fishery from Its Earliest Inception to the Year 1876* (New York: Argosy-Antiquarian, 1964 [1878]), and Reginald B. Hegarty, *Returns of Whaling Vessels Sailing from American Ports* (New Bedford, Mass.: Old Dartmouth Historical Society, 1959). Values as computed in Robert Owen Decker, *Whaling Industry of New London* (York, Pa.: Liberty Cap Books, 1973), and for last three voyages from average price scales given by Decker, *ibid.*

Vessel	Sailed	Home Port	Master	Agents	Destination
Franklin[a]	1 Sept. '59	New London	Edwin Church	R. H. Chapell	Desolation
Roswell King[b]	23 Aug. '64	New London	Robert Glass	R. H. Chapell	Desolation
	13 July '67	New London	J. Church	R. H. Chapell	Heard Is.
	29 June '70	New London	Joseph J. Fuller	Williams & Haven Co.	Desolation
	5 Aug. '73	New London	Joseph J. Fuller	Williams & Haven Co.	Heard Is.
	29 June '75	New London	Joseph J. Fuller	Williams & Haven Co.	Desolation
	28 Aug. '77	New London	Joseph J. Fuller	Haven & Williams Co.	Desolation
Pilot's Bride[c]	27 Apr. '80	New London	Joseph J. Fuller	C. A. Williams	Desolation
Francis Allyn[d]	20 May '82	New London	Robert Glass	———	[Rescued Fuller from Desolation]
	15 Aug. '83	New London	Joseph J. Fuller	C. A. Williams	Desolation
	11 Sept. '84	New London	Joseph J. Fuller	C. A. Williams	Desolation
	22 June '86	New London	Joseph J. Fuller	C. A. Williams	Desolation
	2 Aug. '87	New London	Joseph J. Fuller	C. A. Williams	Desolation
	30 Aug. '90	New Bedford	Joseph J. Fuller	Thomas Luce	Atlantic
	23 Sept. '91	New Bedford	Joseph J. Fuller	Thomas Luce	Atlantic
	6 June '93	New Bedford	Joseph J. Fuller	Thomas Luce	Atlantic

Vessel	Arrived	Sperm Oil (Barrels)	Whale Oil (Barrels)	Whalebone (Pounds)	Sperm Oil (Dollars)	Whale Oil (Dollars)	Whalebone (Dollars)	Total Value (Dollars)
Franklin[a]	No report							
Roswell King[b]	30 Apr. '67	11	No report	645	788	16,387	765	17,934[e]
	19 May '70	—	703	3,223		12,895	2,739	15,634[f]
	26 Apr. '73	—	602	—		12,363	—	12,363[g]
	29 Apr. '75	30	633	1,800	1,521	15,356	2,232	19,110
	18 May '77	25	750	2,989	889	15,561	7,472	23,922
	? June '79	—	950	—			—	—[h]
Pilot's Bride[c]	wrecked		No report					
Francis Allyn[d]	17 Apr. '83	—	No report	—		[8,165]	—	[8,165][i]
	6 July '84	—	[480]	—		10,584	—	10,584
	21 July '85	—	600	—		5,670	—	5,670
	5 May '87	—	400	—		6,552	—	6,552
	3 Apr. '89	—	650	—		4,668	—	4,668
	22 Aug. '91	635	390	—	13,801		—	13,801
	31 May '93	—	—	—		8,804	—	8,804
	2 Aug. '95	—	650	—		14,377	—	14,377

a Schooner, 119 tons.
b Bark, 134 tons.
c Schooner, 194 tons.
d Schooner, 106 tons.
e Sent home 1550 bbls. whale oil, & 4000 lb. bone.
f Sent home 1550 bbls. whale oil.
g Sent home 1,750 bbls. whale oil, 5000 lb. bone.
h Wrecked on Desolation.
i Share claimed by Fuller.

APPENDIX 2

WHALEMAN'S SHIPPING PAPER, SCHOONER *PILOT'S BRIDE*

It is agreed between the Owners, Master, Officers, Seamen and Mariners of the Schr. *Pilot's Bride* of New London, Jos. J. Fuller, Master, or whoever else may go Master, now bound on a Whaling Voyage, from the Port of New London to Desolation Island.

That, in consideration of the shares affixed to our names, we the said Officers, Seamen and Mariners, shall and will perform the above mentioned voyage, promising hereby to obey all the lawful commands of the said Master, or other Officers of said Ship, and faithfully do and perform the duty of Seamen and Mariners as required by said Master or Officers, by night or by day, on board the Ship or in her boats, and at all places where the said Vessel shall put in or anchor at, during the said voyage, to do their best endeavors for the preservation of the said vessel and her cargo, and on no account or pretence whatever, to go on shore or on board of any other Vessel, without leave first obtained from the Master or commanding officer of said Ship, hereby engaging that forty-eight hours absence, without such leave, shall be deemed a total desertion, and in default of the performance of their duty in any particular, they severally agree to subject themselves to all the penalties and forfeitures of the several Acts of Congress, for the government and regulation of Seamen in the Merchant service. And it is further agreed, that in case of disobedience, neglect, pillage, embezzlement, quarelling [*sic*], habitual intemperance, desertion, or the sale of a share or any part thereof before the return of said Vessel, the said Mariners do forfeit their shares, together with all their goods, chattels, &c., on board said ship, or in any store or place where they may be lodged, to the use of the owner of said Vessel, and moreover shall be liable to pay all damages that may be sustained by being obliged to hire other Seamen or Mariners in their place. And it is further agreed by both parties, that each and every lawful command that the said Master or commanding officer shall think necessary to issue for the effectual government of said Vessel, suppressing immorality and vice of all kinds, shall be strictly complied with, under the penalty of the persons disobeying, forfeiting their share, together with everything belonging to them on board said Vessel. Hereby for themselves, heirs, executors, and administrators, renouncing all right and title to the same. And they also further agree, that they will each of them perform twenty day's labor in fitting, loading and unloading said Vessel, (and in default thereof to pay for the same). And the owners of said Schr. *Pilot's Bride* hereby promise, upon the fulfilment of the above conditions, to pay the shares of the nett [sic] proceeds of all that shall be obtained by the crew, during said voyage, agreeably to the shares set against their respective names, as soon after the return of the voyage, as the Oil or whatever else may be obtained, can be sold and the voyage made up by the Owners or Agents of said Ship, first deducting all such sums as may be due from them to the owners or officers thereof, for advances, supplies, or debts arising from any other consideration. And we hereby pledge to the owners, our several shares as security for the payment of the same. It is also further agreed between the owners of said Schr. *Pilot's Bride* on the one part, and the Captain, Officers and Crew on the other part, that if the Captain, Officers and Crew, or either of them, are prevented by sickness or any other cause, from performing their duty during the whole of said voyage in said Schr. *Pilot's Bride* that any of them so falling short, shall receive of their lay or share in proportion, as the time served or duty performed by them is to the whole time said Ship is performing her voyage. And it is hereby understood and mutually agreed, by and between the parties aforesaid, that they the said Officers, Seamen and Mariners, will render themselves on board the said Schr. *Pilot's Bride* whenever thereunto requested by the Owners, Agents, or Officers of the said Ship. And they further agree to be subject to the usuages [sic] and custom of the Port of New London in reference to this agreement.

IN TESTIMONEY [sic] OF OUR FREE ASSENT, CONSENT AND AGREEMENT to the premises, we have hereunto set our hands at New London the day and date affixed to our names, agreeing to pay Interest & Insurance on our advances at the rate of 15% per ann.

SOURCE: Whaleman's shipping paper, Federal Archives and Record Center, Waltham, Mass., Records of Collection District of New London, Shipping Articles, Box 87b. Names of the vessel, port, master, and destination were left blank to be filled in on the form in ink. The final clause, "agreeing to pay Interest . . . ," was also added by hand.

APPENDIX 3

KERGUELEN ISLAND — ISLE OF DESOLATION
by Capt. Joseph J. Fuller

In this short and graphic description of the above named place, I will endeavour to give the reader as good an idea of the place and its surroundings as possible. The island proper lies in 40°41' S. Lat., 69°04' E. Lon. This place and the surrounding groups — including Heard Island, five degrees more to the southerd — is the only land that has been visited by man to the southerd of the continent of Africa and lying in the Antarctic Circle.[1] In my description of the island I will also describe the surrounding groups of islands, they being some of the extreme points about the place. The formation and general appearance of the surrounding groups is identically the same as that of the island proper.

The bearings of the surrounding islands are to wit: Bligh's Cap, lying in 48°29' S. Lat., [the] extreme northerly island. The most southerly point of the mainland is Cape Challenger, lying in 49°43' S. Lat. To the southerd and westerd we have Solitary Island, lying in 49°50' S. Lat. The most westerly point is Cape Louis on the mainland 68°38' E. Lon. The most easterly point is Cape Sandwich, lying in 70°34' E. The most northerly [point] on the mainland is Cape Français, lying in 48°39' N. Lat.; and the most westerly island we have is Fortune Island, lying in 68.32 E. Lon.

Approaching the island from the notherd and westerd side, the most convenient harbor of egress is Christmas Harbor. As we make for the harbor from the N'd. and W'd. we pass by Bligh's Cap. This, like the rest, is uninhabited and very rarely visited by man. Sea elephant, leopard seal, and birds abound here. Formerly was quite a place for fur seal, but of late years it has been so frequently visited that the seal have left.[2] Approaching the above point from the westerd, we come in close approximately with the Cloudy Isles. They are two in number, the most northerly one being the largest. They have the same characteristic appearance as the others. They are studded with mountains of medium size. Here and there can be seen valleylike openings extending towards the interior. There is a channel extending along between the main and [the] other island. It abounds with reefs and is very dangerous at times in foul weather.

We at last arrive at Christmas Harbor. We enter it and if there happens to be a strong breeze or gale prevailing from any of the quarters but the easterd, you will lie as steady as if you were in a mill pond. Arch Point on your right, departing, is a narrow strip of land that serves as a breakwater against the sea, and Cape Français serves the same purpose on the n'd and westerd. After having entered the harbor, you take in your surroundings. Facing the westerd can be seen the two mountains, Havingal and Table. The former is 1,130 feet and the latter 1,275 feet above the level of the sea. Between these two mountains is a valley, vegetation growing very abundant. At the base of them there is a beach extending inland for . . . [no figure given]. Vegetation makes its appearance on the sides, but gradually disappears towards the summit and scantily growing about may be found a bit of moss, but generally summits of these mountains on the island are invariably barren.

This bay was first entered by Kerguelen, the French explorer. It being on a Christmas day, he named it after that day and gave the island his own name. The outcome of his discovery I give the reader a full account [of] this in the forepart of the book. From here we go down around Cape Cumberland and head to the westerd up Cumberland Bay. It extends for some nine miles. There is a certain kind of picturesqueness about the shores as you go up. The Bee Hives are a small range of mountains extending for about six or eight miles and resembling a cone of honey very much. Saddle Hill is also quite a prominent place, resembling a saddle very much.

Coal is to be found towards the interior of this portion of the island, being about of the same quality as soft English coal, making very good fuel. Of course they have never utilized any of it. I[t] was thought of one time, by the British government to establish a coaling up station here for ships bound for the Antipodes, but the project never met with favor. I believe myself that it would have been a source of wealth to them as it would be an easy matter for the mines to be opened. The stuff is very abundant here; in fact, it is to be met with in all parts of the island, very near.

In the vicinity [of] Christmas Harbor can be found the remnants of the good sized trees in the state of petrification. They must have grown on this island at one time, as they are imbedded in the soil for nine [or] ten feet, this being conclusive proof that they could not have drifted here, as even the twigs and very fragile branches are to seen. A strata is formed over it. This being the case they couldn't have come from other parts. You can see the grain on the petrification and it has a slateish appearance. Of course, no specimens of gigantic size are found in this portion of [the] island. I have come across limbs that would measure in circumference five or six feet and four times as long.

First of all and before enter[ing] into details about the place and for the convenience of the reader, I will say something in regard to the vegetation of the island. Everything that grows here with the exception of one plant, that being a kind of sorrel, having been imported here I suppose and being taken ashore by someone, it grows spontaneously and I might say with luxury as it seems to be as hardy here as in its native clime. About the most plentiful plant that grows here is the Desolation tea, a species of rose.[3] It makes a very wholesome drinking, having a very pleasant and agreeable taste. It is of an acidious flavor. It grows rank wherever the soil will give it a footing. The roots are imbedded very deep in the soil and are of a brown color. They are very long and of about [the] thickness of a large finger. They are not fit to eat. They spread out for a good distance. The plant bears a flower resembling the fruit of the strawberry very much, of medium size, of a crimson scarlet color, having small thorny projectiles protruding out of them. If any fabric comes in contact with it, it will stick to if like a burr. It is about the size of a 10¢ piece. This plant or herb, as you might term it, grows the most plentiful on the island.

The next we have in abundance is Desolation cabbage.[4] This is met with very frequently on the island and is the only thing in the vegetable line on this place. In appearance it resembles our article at home, having the characteristic leaves and the same root. It does not attain the size or does not have the flavor. I do not know of anything that I could possibly compare it with as regards flavor. It has got a slight bitter taste, which I am sure could be removed by a little cultivation. I myself, and I have heard other persons express their opinion on it, prefer it to the best cultivated cabbage. Its interior leaves are the same as those of the cultivated sort, while in color [they are] very tender and more savorious than the exterior leaves. It is sought for by all persons coming to these parts and is considered a stable article. It makes delicious cold slough [that is, cole slaw] and I have no doubt but what saurkraut could be made out of it. In all, I am certain that it can be turned to all the purposes that our cultivated kind is.

Suffice to say that it is quite a boon to whalemen and others coming to these parts. Even — I will add — the transit of Venus party seemed to consume an enormous lot of it and I can assure you that they relished it, as I have seen them especially send their steam launch off to gather it and they raised havoc among that year's crop. I have made an attempt to import some of it into the States this voyage. I find myself off of the islands of Bermudas and I see that it [is] putting in a withered appearance. Owing to the hardiness of the plant, I do not really think that the temperature has affected it any, as it was subject to every bit as warm weather in Desolation as what it has been this passage. So it must either be in the water or atmosphere. I have also got some of the tea. But I think it will be a failure like the cabbage.

The next we have in abundance is a species of moss resembling a tussock very much.[5] Its stem is reedlike and about three inches thick. Its roots grow very long but are not imbedded in the soil very deep, as with the slightest pulling they will come out. They attain the thickness of a man's arm and are very good to eat, having a flavor very much like beet root. They grow in clusters and on the sides of rocks. This plant grows in the most desolate and arid part of the island. The fern also exists here. The similarity of it is very much like to that of ferns of other countries. It bears the characteristic tassel from which the seeds come. It does not grow very luxuriantly and seems to be stunted, as it does not attain the size that it does in other more temperate climes.

There is a flower that grows here. I have no knowledge of its name and do not know if it is a native to this place or not. It grows about three inches high and bears a sky blue flower. Its foliage is dark green and does not grow very rank. It has no odor and is of no account but for food for cattle. It is of a scarcity on the island. There is also a kind of grass on the place. It grows quite abundant in tufts and patches, and is most plentifully to be found where there has been some habitation, especially around old vacated houses. It is about the same as our grass at home, only it does not grow so tall. The fungi — mushrooms and toad-stools — put in their appearance and seem to thrive. There is the two varieties of them, nonpoisonous and poisonous. The soil being very rich here, it grows almost any place. Of course I think it has its season of growth.

Now, kind reader, I have given you a botanical description of the vegetation of the place and I will now give you information as to the etymological order. We have no venomous insects here. I have met with two subjects of the spider family. First, the old familiar daddy longlegs — of nursery fame — and, then, a kind of black spider, about the size of an ordinary water spider. These two above insects are quite abundant on the island. They principally inhabit the crevices in the cliffs and are to be met with in the grass and foliage of the different plants. I suppose they live on flies or other minute insects.

We have a native fly the same size as an ordinary house fly; in fact, its appearance in every respect is like that fly.[6] They are most commonly seen amongst the decayed vegetable matter. My attention was first drawn to them amongst the cabbage. They are not a pest as they never invade the lodgings of man. They are very lively on their feet and hop about or rather spring about from place to place. They do not seem to take to animal matter, as I have never seen them. These are the only insects that I know of — not a large variety, you see.

I will now say something in regard to the piscatorial species: [first] the bivalves and crustaceans. There is a clam here, of the soft-shell kind. They are called long necked clams in certain portions of America, only their shell is more soft. And there is also to be found the remnants of another species on the west side at Kerguelen's Head. They are in the state of petrification and look like as if they had undergone the process of fire, as they are found in large lumps adhering together, making them appear as if they had been hove up by some volcanic eruption. When broken open the interior, which was formerly the meat, drops out and looks like a burnt mass. The slabs that you find look like a charred mass of shells. The former ones are very good eating. We have two varieties of mussels. Both are delicious eating, the bog mussel and [the] black, the ones with the black shells being the most palatable. The others when opened, the inner shell has a bright mother-of-pearl look. The sea snail also inhabits these parts; they are one of the slug kind. I have picked them up on the lee side of the island. The way we procure the mussel is by going in quest of them after the tide has gone out, and they generally can be found in any abundance.

We have quite a number of fish about. The most remarkable one is the devil fish.[7] It gets this name from presenting such an ugly appearance. The head is of enormous size in comparison to the body. In fact they are all head; from the neck, the body gradually tapers down to a point. The head is adorned with prongs similar to those of an ordinary cat fish, and their mouth and eyes are a marvel for longness. The color of this fish is dark slate and will weigh [a] pound or more. They are very sluggish in their habits, very slow swimmers. They are best caught in the harbors in shallow water.

The longest eatable fish there is here is a fish resembling our common cod very much, the head being a bit longer, and known down here as the "big fish." They will weigh from ten to fifteen pounds and are splendid eating. Two species of the fan fish abound here and are known as "night fish" in these parts. They are very good eating, their color being variegated, and will average a pound or a pound and a half apiece. They are very plentiful, being caught by the bucketfull and are thickest among the kelp — seaweed. There is a small fish called the rockfish. They are excellent eating. The way that they are procured is by going along the beach after the tide has gone out and looking for them under the rocks, where they are to be found in large numbers.

If the reader now wishes to obtain any further information as regards the different things to the island, such as seals, elephants, birds, etc., etc., he will please refer to [the] pages [below]. I will now proceed to give a general description of the island. Its resources, vegetation, and inhabitants of different kinds I have already spoke of, so I won't further dwell on them but will commence with the geological formation of the place. On the island, strewn about quite profusely is a stone belonging to the first strata of the crust of the earth. It is the basalt of antediluvian origin. Croppings out of it are to be seen and also ledges. It can be found about the island where it has broke off at the main body. This goes to prove that there has not been the change enacted on the island that there has been on other bodies of land. I have conclusive proof that the place has been in a volcanic state at one time, as there is old craters yet to be seen on the place and there is one or two of them active. For instance, take the Crozet group. The formation appears to be entirely newer and of a later period than that at this place. The main body of the island is composed of hard stone similar to that that forms our land at home and ice with, of course, a crust of loamy sand or dirt. The latter is very fertile but not claylike, as it will crumble into pulverized condition.

I will describe the icy element first. Glaciers abound about the place and are most numerous on the western portion of the island, as this is the weather side and consequently the most frigid. We will go on a tour of inspection down the western side of the island. Our anchor is hove up and we get underway out of Christmas Harbor and after rounding Cape Français we go on down the coast heading S.S.W. as we will sail along, keeping our gaze on the coast all the while. We see mountainous and altogether desolate looking country before, as on the west side. The vegetation is very scanty and can not be perceived any ways off from shore. On this side of the island, we do not see any remarkably high mountains [on] the place; they are more like enormous big ridges. They extend from the beach, generally gradually sloping from very near the water's edge and at times the ascent is abrupt enough to deserve the name of a precipice. Between these mountainlike formations are valleys invariably with rivulets flowing down them, having their source and being fed by some glacier. In these valleys in most parts of the island vegetation grows very luxuriantly, but not so much on this side as the cold is more intense.

We move on down the coast and the land gradually gets higher. When we get about with Centre Bay we come upon a regular coast range of mountains. This chain extends from the northerd to the southerd. Back of the range is an enormous long iceberg or glacier. It terminates at Centre Bay as it is visible from that place. It gradually increases in size as it extends to the southerd. As we continue our journey down the coast we still see mountains and the valleys extending up through them. Instead of going [around] West Island we make for Marienne Straits. There being a fine breeze prevailing from the northerd and westerd, we won't experience any difficulty in getting through.

Coming in view of Thunder Harbor, we are startled by a loud thunderous noise, which sounds like as if a thousand of cannons had fired a volley at once. As we get a clear view of the place, we find out the cause of it. Enormous large pieces of ice have become disengaged from the main body and have been precipitated down the glacier's side. Icicles weighing no less than several millions of pounds break off and roll down, causing this unearthly roaring sound and creating a tidal wave, as it were. On the sides of the mountains about the place the ground has the appearance of having been at one time occupied by ice. There are thin ridges running up and down the sides and the general appearance of the place denotes that it at one time was covered over with a layer of ice. The way that I account for it is that the ice on the sides of the mountains has formed an avalanche and slid off of the mountainside. There is still ice from the base of the mountains, projecting towards the water. From the water there is another formation. It can be seen cropping out on the top of the beach so it must form on the bottom. I am convinced that it is not so cold here now as what it has been in former days. Scantily growing on the sides of the mountains can be seen the mosses and rarely the cabbage and tea. The summits are quite bare and void of all vegetation.

We sail out of sight of this very solitary sight and get into the Straits of Marienne. We pass Monument and Duncan's coves on our starboard and get well into the straits. This place is considered to be one of the most dangerous and hardest places to navigate in that there is about the island. It is a very bad place to be during a blow from any quarter but the n'd and s'd. The narrowest place in it is about one hundred yards, so you can see that there is not much sea room to move about in.

We move on towards the southwestern shore and we get a view of Young William's Bay in the distance, and are right abreast of Bull and Melissas beaches. Both places are noted for the elephant, Bull Beach, I might say, being [more] famous. Even up the coast there is two places that are very good for elephant, Blue Skin and Shoe Foot [beaches]. The iceberg is about six miles from Bull Beach. It can be readily seen, and I think it extends down to the beach. In places, you can see where the sand has been blown off the ice and the ice lies exposed. In other places you can see vegetation growing on the ice wherever there is a footing for it in the soil. Somewhere in the neighborhood of Bull Beach and Young William's Bay there are hot springs.

We go on down the coast and pass places that are literally swarming with elephant, the names of both of them being of the infernal spots. In rotation the first one we pass is the Devil's Cowyard and below in a kind of bay is Hell's Gate.[8] After passing this place we dare not hug the land so close as we did coming down, as the coast on down to Cape Bourbon abounds in reefs and a heavy surf running in at all times. We get on down towards Bonfire Beach. These three places are inaccessible as you cannot work them. Hence the elephant breed there in millions. I have been ashore in these places myself and have actually seen them lying two deep, that is one layer over another. They serve as a supply for the whole surrounding island. This is the reason that I ascribe for them being so plentiful up at the other west side beaches. They are never troubled in these secluded haunts.

We have now arrived off of Cape Bourbon. Around on the southern and also on the western sides can be seen trees of gigantic size strewn along both beaches quite profusely. They look like as if they had been precipitated down the mountainside as they lie with their roots in a straight position toward the beach. The same thing can be seen on the other side of Cape Bourbon, but these do not lie on the beach in the same position. The trees will measure some feet in circumference and as much as forty feet in length. They are in a fair state of preservation, but, little of the wood being decayed, numerous very small and fragile branches can be found and it resembles the wood that is found at Christmas Harbor in a petrified state.

Now comes the very puzzling interrogative? Whence could these trees have come from? Are they natives of the island? Or have they drifted here with the current from some unknown part? If they are indigenous to the place, I have an idea that they would not be in the place that they are, rather [they would be] on the leeward side, as the wind is so tempestuous on this, the weather side, that it would be an impossibility for trees of any dimension to exist as the wind has even an ill-effect upon the scant vegetation. And furthermore there is no signs on the mountains of them having grown there. So if we cannot account for them this way, I will have to advance another theory and it is one that I am inclined to believe. But still there is something to oppose it. They have been thrown for some fifty feet on [to] the beach, and lie some hundred yards from the water's edge, so the reader can think for himself what an enormous big sea it took to do it. I have never witnessed any thing that could have half-way equalled it. This is my theory on the matter: that they must have drifted over from Cape Horn; it must have taken years for them to do so as I do not think the current is very swift from there [to] here. I will leave the reader to himself to accept either theory. One thing sure, it must be one of the two.[9]

There can always be seen old ships' wreckage about the place. And up at African Bay there was a figurehead found. It is supposed that it came off the *Grand Turk* as the figurehead was that of a Turk, and it's said she carried that kind of a figurehead.[10] She has been missing for some time, but no word has been got of her. I have picked up old burnt spars and planking. In fact, debris from ships of every kind comes ashore on this place. I was taking a walk from Christmas Harbor to West Shore and as I and one of the ship's hands were walking leisurely down the beach I espied something in the water. I could see it was a bottle, so I went over and fished it out and found upon looking at it that it was a two-quart bottle of best whiskey; so we took it along to the vessel with us. I have found paraffin wax, dried apples, cracked peas, all of which were in good condition and eatable. These ships must have been bound for the Antipodes. Vessels have been lost here abouts and no word or tidings ever got of them.

Cape Bourbon is [the] next extreme southerly point [next] to Cape

Challenger. It is here that the discoverer Kerguelen — ill-fated as he was — put an end to his career. The details in regard to this will be found in the latter part of the work. The glacier that extends down this coast terminates within about ten miles of the extreme point. This unoccupied land is level and gradually makes an ascent. It is around on the western side of this cape that can be seen a volcano, or, I dare say, a geyser. When the water rushes in, it will spout up large clouds of steam; when the surf recedes, it will calm down, to repeat it again when the surf goes in.

We will now take a glimpse of the southern portion of the island. The first place that we come in view of is Little Half-Moon Beach. This is a noted place for elephant. The next, White-Ash Beach, is also a great place for elephant. In fact I might say all of the harbors and beaches on this side are accessible and abound with elephant. The land hereabouts is very rugged and very broken, but there is not so much ice to be seen as on the western side as we sail along [past] Sprightly [Bay and] Table Bay. [The latter] gets this name from the fact that there is a mountain there that resembles a table very much. We come in view of Mount Ross. This is quite a prominent mountain here on account of it being the highest on the island. It towers 6,120 feet above the level of the sea.[11] Its summit at times is away above the clouds, and perpetual snow exists there. The glacier extends to its base and then extends on to the westerd. Passing Half-Moon Beach, we come with Swain's Bay. Swain's Bay offers the best kind of protection. There are Shoal Water Bay and Volage Bay.

There are some mineral springs to the northerd of this place. While the American transit of Venus party was down here, a Professor Kidder requested me to procure some of the water for him.[12] While up at Shoal Water, [I] complied with his request and got a bottle full of it. But unfortunately the bottle got broke, so he did not get to analyse it. I have tasted it myself and [it] acts upon your mouth in the same manner as the crystallized alum. Whether there is any medicinal properties attached to the water I do not know, and I have never seen it tested. These hot springs are quite numerous on the island. The water oozes out of the ground and is very warm.

The appearance of this part of the island varies but little from that of the western portion, only I think that there is more vegetation to be seen. The country still has that rugged wild appearance. We make our way on down around Cape George and pass by Greenland Harbor. At last we arrive at the entrance of Royal Sound. This harbor, as I might call it, extends for twenty-four miles and is about twelve miles [wide]. It forms a regular archipelago. It is full of numerous little islands of no importance, as no amount of sea elephant are ever found on them — only a few shedding ones. These places afford excellent shelter for a vessel. The surrounding land is all very high. Good sized mountains can be seen. Even the islands all [are] very high and mountainous land. Malloy Point lies in this place and it's here that the American party established themselves to take the transit of Venus.

We leave here and sail around the Prince of Wales Foreland and head on up the coast. After leaving Shoal Water Bay, we come in view of a country entirely different in appearance from that of any other part of the island. It's about the most level and unbroken land on the island. There are only two or three mountains on it, the most lofty one being [Mount] Pepper, 650 ft. There is a good sized lake here, fresh water. It is some thirteen miles around it. It has its outlet in a stream that leads to the sea. Near this there is a remarkable looking green mound. It is a good sized mount [or] knoll, with a verdant cover of vegetation on it. It can be seen from a good way off and is quite a contrast to the surrounding country.

We round Cape Digby and head back for Christmas Harbor.[13] The appearance of the land is the same as usual. It grows rugged and mountainous. We pass Betsy Harbor, better known as Pot Harbor. We now come to where the islands are very numerous. The bay, harbors, and sounds are literally full of them, Whale Bay, Rhodes Bay, White Bay, Centre Bay, Bear-Up Bay, and Irish Bay being the most important ones. We go on between Swain's Islands and Howes Island, both places being of importance in the way of fur seal and elephant. We at last arrive at Christmas Harbor. As I have said, it was here that Captain Cook first landed. He made this place his headquarters during his stay. After leaving here, he went into Port Palliser. One of his men must have died on the way there, as he buried one on this place. The grave is yet to be seen, but the head board has become so blurred that the characters can not be discerned on it, so you can't possibly tell who it could have been. [On] Winter's Island, a small island in Whale Bay, Goldsboro the explorer wintered when down here.[14] There are many haul-overs about the island and many miles can be saved at times when traveling in a small boat by hauling the boat across them. They average an eighth of a mile in breadth. The island is now explained and the reader can draw his own impressions from the description given of it, the continuation of the book, and the different things about the island.

THE KING PENGUIN [OR] GREAT PENGUIN

This bird[15] is the largest of that species and, I dare say, the most prepossessing and handsome. The other varieties are mere pigmies alongside of this bird. When full grown he will weigh fifty pounds and possibly more, and will measure three feet and a fraction. The appearance of the bird at a distance resembles a human being very much, as I have known men to confound them with men that were absent and were expected back. They would see a couple of penguins coming down a hill and at first they would be sure that it was the absented parties. They would approach them only to find out the deception upon approaching them.

They are a remarkable bird. If you happen to be walking along the beach in the vicinity of where their rookery is, you will come upon a foot path grooved down into the earth for several feet, and then you will see a whole brigade of kings marching up and down it. Constant wear for years and the effect of the rain makes a regular pathway. Following this pathway inland for some miles, you come in view of their solitary haunts. Here they are properly in their glory. Thousands upon thousands of them can be seen, and I must [say] that it's a grand sight for the human eye to witness. Standing upright and strutting about with an air of importance, with their militarylike coat, gives them the appearance of a soldier doing duty. They march about in files and by twos. Some of them have the part of sentries. As I have said, they are a very handsome bird. To give you an idea of what an interesting sight it is to see them in their natural state, I will relate an incident that happened here, during one of my first voyages as master.

It happened that amongst my crew I had a young German. It appears that his father was a large mill owner and manufacturer of cotton and woolen fabrics. He sent him over to the States to make inquiries and to investigate some improvements that had been enacted in that business. He arrived in New York, and went around to see the tiger. He commenced to dissipate and squander his money right and left, until he found himself penniless and without friends. It was the same old story: plenty of associates and friends as long as he had money, but alas and alack down [with] him after he became destitute. Being without a place to stop or eat, and being ashamed of his conduct, not wanting to let his people know, he determined to go to sea. He presented himself to Morrison, the New York shipping master, who duly crimped him and he was sent down to New London and was put aboard of my vessel bound for Desolation. He performed the voyage with credit, seemed to be very eager to learn, and was willing and industrious at all times. By the time we arrived here he had developed into a full-fledged sailor.

It so happened that we were ashore one day and were going over to the pinnacle and one of the largest rookeries on the island was situated here. There was a kind of bluff in front of it, so it was totally obscured from us until we broke in full view of it. The German was thunderstruck with what came in view, so much so that he stood in wonderment and admiration for several seconds. At last he gave vent to his intense astonishment. "Holy Mans," he blustered out, this was one grand and magnificent sight. "Mine eyes have never seen the like before. Here I have come all the way from Yermeny, and what for. I tell you, captain, that this is one grand sight. And I am satisfied to return to my Faderland." So spoke Hans. And I can assure you that he meant it. To tell you the candid truth it is one of the grandest sights that I have ever witnessed, and I am sure that there are many of you that would evince as much interest as what Hans did over the matter.

I might as well give you further details in regard to this young German. When we left here we went by the way of the Cape of Good Hope and called in at Cape Town. When I got the mail I found that our friend Hans had a couple and after reading I came to find that his father had forwarded him a draft, it being a sufficient amount to take him home and it came in my care. He begged me to discharge his son, etc. I asked Hans if he wished to leave and he said no, that he had his mind made up to perform the whole voyage. And he did so with entire satisfaction to all.

The bird's plumage is very soft and downy around back and belly. The wing feathers are very stubby and more down than anything. Their backs are a glossy black hue and their breast an alabaster white. They have very straight bills and from the sides of each eye there is a streak about two inches broad [that] extends down their shoulders and gradually disappears on their stomachs. This stripe is of variegated color, bright yellow predominating, and looks very much like a regalia that is worn by different society. This adds more to their soldieringlike appearance. Their feet are jet black. [Their] tail is inclined to be stubby and the same color as the body.

For their rookeries they always choose a place where there is an incline, as they are strict believers in sanitation. This idea in selecting ground where there is an incline to it is so that the refuse, etc., will be washed away. They do not have any nests; that is, they have none constructed out of grass sticks or any other material. A small pouch between the legs in the region of the stomach serves as a receptacle for their egg and the young, when hatched, occupy the same position. From accounts got out about this island by several very distinguished gentlemen, amongst them a Doctor Kidder, I see that they kind of misbelieved my assertion at home, although I am certain that the professor was convinced of it. He could hardly help from believing it as it was verified by everybody that had ever visited that rookery.[16]

Now here is another very absurd report that certain travelers have got up about the penguin and albatross: that the albatross makes his nest, or rather lays his egg, with the penguin, and this bird does the incubating. I have studied into the subject pretty close and I can't see how this belief ever got current. Both birds have distinct rookeries and some miles distant from each other and they never intermix. And it could not be the king [penguin] because they have no nests. I think myself that this report was got up more for sensation than anything else. I have been in and about these South Sea islands for scores of years and I know men that have also been cruising to them for years and they do not know of any such a thing — only knowing of it in print.[17]

If the bird is with egg, of course, her impregnance will make her walk very awkwardly. If you want to get the egg, put your hand under their tail and the egg will slip out of the bag. They can be made to lay several times by robbing them. I have often seen a penguin walking along with the head of a small one sticking out from behind and looking as contented and comfortable as if it were acting under its mother's wing. They lay in the month of November. I suppose that they first put in their appearance in the bag. The [egg] is hatched there. They are very cute and comical when they are young and especially in appearance. They resemble an old and venerable gentleman with [a] fur jacket on.

They are very palatable when young and even the old ones are splendid eating. In fact they are the best and most wholesome bird for eating purposes that there is on the island. They are very oily, having a layer of fat underneath the skin. They have no craw, but digest their food like other aquatic birds. They can be drove like a parcel of sheep or swine. When landing they always do so about a mile or so from their rookery. This, I think, to escape the leopard seal, as he feasts on them whenever he gets an opportunity. He is aware where their rookery is and he lies in wait for them. They land away from it to elude him. They attain their food in the [sea], it being squid and fish of a small size. I have seen them drink fresh water when heated, but as a rule I do not think that they partake of much of it.

As I have said, these birds are very tame. I remember one time I was seated, taking a rest after a long walk on the beach, when a number of king penguins congregated around me. One of them was extra bald-headed. He was a peculiar looking fellow. He had a remarkable looking head; he had a good sized [top] knot on it of the same color as is his back. Now, it is a rare thing to see one of these fellows, [and] this fellow was making himself very sociable with me. I offered him a piece of pork and [he] greedily took it and swallowed it down. I had a mind at the time to capture the fellow and take him aboard, but as it was rather too early in the season I thought it best to wait. But at the end of the season I could not find one like him, so I did not bring one along with me.

THE MACARONI AND ROCKHOPPER PENGUINS

The macaroni is of the same size as the Johnnie, being about one or two feet in height. Their heads are adorned with white spots of [the] size of [a] 10¢ piece that lead over the eyes, only in color they are of a lighter shade. The color of their body is the same as that of the other penguins.

Their legs are yellow and their feet pink. They take to broken-up country and crevices on the low lands, although I saw a remarkable freak in their habits at Cape Bourbon, where there [is] a rookery about a mile square and it is built on a perfect plane. This is nowhere else to be seen about the island as they invariably take to rugged and broken-up land for their haunts. Their nests are composed of sticks and grass. In this they deposit two or three eggs, somewhat larger than those of a hen. Of course, they can be made to lay for months if they are robbed. The eggs and the bird are both very wholesome eating. They have a weakness for thieving, as they will steal eggs from one another when ever an opportunity offers itself. When disturbed they are very vicious, as they will fly at you in a very angry manner. They are met with four and five hundred miles from shore.

The rockhopper is the midget of the whole family. He not only differs in size but also in appearance. Like the rest, the predominating colors are black and [the] white of [the] topknot [on] his head. They inhabit the rocks and fallen cliffs, hence their name, rockhopper. The habits of these birds are identically the same as others. They lay two or three eggs and can be made to lay more if robbed. Like the others, their outward appearance does not distinguish their sex. They commence to lay in . . . and the brood puts in its appearance in. . . .[18] They also are quite fierce when disturbed.

THE JOHNNIE PENGUIN

This penguin is the same size as the macaroni. The predominating colors on its body are velvety black on its back and snow white on the belly. Its beak and head are the same as those of any penguin only it has while spots of the size of a 10¢ piece on each side of their heads just around the region of the eye. They are very plentiful here. They lay an egg very much the same as that of the king only smaller. They can be made to lay in five or six months. This is done by robbing the nests. Their egg will gradually decrease in size and last ones will be void of the yolk and will contain nothing but the white. When you approach them the ones on the nests won't make any attempt to flee. You take their eggs from them. If you leave them near the nests they will start to roll them back again. I have often seen them steal eggs from one another. When you depart from the rookery they will set up a humanlike laugh. It resembles that of a person very much. It is the ha! ha!, and can be heard at a good distance.

Their rookeries are generally situated amongst ridges and hills and [they] construct their nests out of grass and moss. When their nests are not molested, they will hatch [two] young. The first-born receives marked caresses and favors from the parent birds. Whenever they arrive with food they will always invariably feed him and what is left they will give to the rest of the brood. They seem to make a marked distinction between him and the others. The young are very frolicsome, as they will go through all kind of antics with each other. They resemble so many children playing at hide and go seek. They will catch a hold of each other and tumble about and then let out a laugh, which makes it a very interesting spectacle to witness.

When the old ones think that their brood is old enough to take to the water, they will get them in file and march them down to the water's edge, as their rookeries are situated some way from the water. Then they will try to entice their young into the water, going into the water first. The young will follow up to the water's edge, then they will rush up on to the dry beach. At last they venture into the water and, not having any previous experience, they will be dashed ashore. They are undaunted at this and seem to enjoy it as they will rush off into the surf again and perhaps meet with better success. In a few words they become so that they are independent of the old folks and can look out for themselves in the waters.

THE SEA ELEPHANT

The sea elephant inhabits those islands lying to the southerd of 37° in the Antarctic Ocean.[19] They are killed in vast numbers by whalemen and others, who go there in quest of oil. In appearance at a casual glance, they resemble a gigantic hair seal, although they are like their brother in many respects both in appearance and mode of living. The male — bull — grows much larger than the female — cow — the latter attaining not more than one-eighth the size, so you see that they are mere pigmies in comparison with size with a bull. The head of a bull — when young — resembles very much that of a bulldog, but when they attain a good age his canine look disappears. Their very short and stubby nose eventually increases in size until it becomes, I might say, a proboscis or trunk, which gives them the appearance of an elephant, hence the derivation of their name sea elephant. The bull, when full grown, will measure in length nineteen to twenty-four [feet] and will measure around his body in the region of his fore flippers from ten to thirteen. A fair sized bull that has just lately arrived out of the water will produce from eight to eight and a half barrels of oil. Each barrel contains thirty one and a half gallons. The longer they remain on the beach, the leaner they get and, of course, won't produce as much as a newly arrived one.

The bulls are the first to arrive. When he first comes out of the water he looks quite sleek and fat. His coat is of a light brown inclined to be shiny, but after exposure it turns to a dark hue. As regards texture and quality, it is of about the same thickness as a bullock's, quite as abundant and, of course, naturally greasy. The flippers on the fore part of the body are seallike and are adorned with five nails on the flipper. They are used for scratching themselves with and I suppose to help procure their food with. The fore flippers serve to steer themselves with. They have very broad and massive breasts. From their shoulders their body commences to taper down. The after part of their anatomy is very much the same as an ordinary fur or hair seal. The tail flippers are very large. They are used for propelling themselves through the water. They work on the same principle as an oar when sculling with it.

But little is known of this animal during its absence from the beach. It arrives in the month of September to have its young and again disappears until the latter part of October and it again arrives to shed its hair [in the] latter part [of] November. The cows, during the month of their confinement, lie in the sun in a kind of semiconscious state. Neither the bulls or the cows are ever seen eating anything. Now comes the question. What comprises their food? What do they subsist [upon]? This question has puzzled all of us. One thing is true, they provide their food in the sea. When getting the blubber off a number of them, I have made it a point to dissect their intestines, but nothing that would give the slightest clue to the above question has ever been found. Their stomach contains a yellowish substance looking very much like yellow paint, but when exposed to the sun it becomes a hard dry substance and turns to a dark amber hue. No particles of bone, scale, or other hard matter can be found in their stomachs. I have found stones of about the size of a goose egg inside of them. Whether they devour these for food, ballast, or what, I cannot say.

They congregate in water holes in large numbers. They seem to like fresh water very much. That they can remain under the water for a good space of time I have seen proven already. One morning I noticed a bull lying in a fresh water hole. I made it a point to return there. That evening I passed by there and he had not budged a foot. Next morning he was in the same condition. His head was completely submerged under the water. Thinking that he might be dead, I approached him and drove him out. At first he did not care about moving but after a little kicking he arose out of the water. I did not further disturb him in his altogether watery bed, but left him lie there. After a little while he again lied down, his head again under the water. When he rose his head out of the water, he did not seem to be short of breath. This is conclusive proof that they can stay under water for any length of time.

Whatever comprises their food it must be of a very soft and easily masticated kind as their teeth, you can plainly see, were not made for any kind of hard substance. They are hollow and are tipped with enamel. I have taken particular attention of the teeth and I have never found any worn out or decayed ones, although I do not know if they are subject to decay. I have killed old bulls who had not a tooth in their head, not even the remains of any, and they were about in as good a condition as any of them. This I think is conclusive proof that whatever composes their food is very easily masticated.

THE SEA LEOPARD

This animal,[20] like its brother the elephant, inhabits the same regions. We know less of this animal than of the sea elephant. They are a great deal smaller, hardly attaining one-eighth the size of a good sized bull, and [I] might say that they average the size of an ordinary cow. Their habits are rather obscure. They are animals of prey, as I have watched them capture penguins and other sea birds and have found the remnants of fish in their stomachs. You will note the anecdote related here on pp. 65–66 in regard to them procuring their food. Their head resembles very much that of a sheep. In general appearance they look like an ordinary hair seal. Their tongue is very much like that of a bovine. It is long and fitted with a rough surface. They have two small holes that serve as ears. Their jaws are very much different in respect to size and appearance to any other seallike animal. Their jaws are studded with pronged teeth, except two on each jaw that are tusks. Their color is a slate blue, and sometimes black. Their coat is adorned with numerous spots, about the size of [a] half-dollar and of dirty white. Little is known of how they breed and rear their young.

THE ALBATROSS

This majestic looking bird truly deserves the title of "The King of Marine Tribes."[21] I myself think that it is the most noble looking aquatic bird there is. They are a beautiful bird. Their plumage around the neck, breast, and belly is of a snow white, shading delicately into grey and dusky brown at the sides and back. The feathers are so curled and elastic that the skin with the plumage on it is an inch or an inch and a half thick. I think myself that it would make excellent stuff for wear on articles of clothing. It is very light, yet exceedingly warm, while for appearance its dovelike smoothness and purity can not be excelled. It has the advantage too of being very durable, the natural oil of the bird preserving the skin and feathers for many years, while the characteristic musky odor is very easily overcome by camphor, although I learn that there is no commercial value attached to the bird.

The albatross is really a king of the marine feathered tribe. He is not a vicious bird, unlike his land-living brother, the eagle, and besides he is not a bird of prey. He procures his food from the sea, it being a kind of flying squid. He also seems to like any kind of salt meat, as I have seen them hover around the ship for days, devouring a bit of pork or salt beef that would be thrown into the sea. They are very keen sighted as they will be several miles away from a vessel and they can tell whenever there is anything eatable for them thrown into the water. They will reel around it for just a moment and then pounce on it, and gobble it down to arise again and to repeat the operation if anything else comes their way that will tempt their appetite. I have also seen them eat the blubber of the whale and eat the flesh of the carcass after it is adrift. Of course, they will only do this in certain cases, when there is I suppose a scarcity of squid.

They are very fleet of wing and are about the best flyers there is. They have been known to fly distances that would sound almost fabulous. During one of my cruises to the Crozets, I had a party of six men stationed at East Island who were sealing. There was a French man-o'-war come in at the time. She did not anchor but laid off and on about the place. They sent a boat off to East Island and then the men noticed it coming over to where they were. They went down to the beach to meet them. Upon landing they asked them if they "Parle voux Francey?" Of course, the men were mystified at first. But it happened that the officer in the boat was conversant in English, so he asked the men who they were. The men told them that they were employed catching seal. And then the Frenchman said that he was in hopes at first that they were the party that they had come to those parts in quest of. They then went on to state that there had been an albatross caught off of the coast of Australia and it bore a copper plate around its neck. There was an inscription on it and it was in the French language. They took it to the French consul and he found that it had been attached to the albatross by some castaway Frenchmen. It gave the bearings of the place where they were and the date of the flight of the bird. Of course he informed the French government of it, and they immediately dispatched a ship after them. The copper plate had stated that they were on Hay Island, but they [had] gone to this place and found it abandoned; but they found a letter there stating they were going to remove to another island, East Island being the one. They cruised in and about the place for several days, but did not learn of the fate of their countrymen. They must have made an attempt to get over to some other islands and got caught in a blow and perished.[22]

There is another instance to illustrate the rapid flight of this bird. The captain of a whaleman, while cruising in these parts, caught one of them and attached a copper plate to his neck. Sixty hours afterwards it was caught off of the coast of . . . [blank left unfilled by Fuller]. I am perfectly convinced myself that they are the most enduring and rapid flyers there is. The bird in his solitary haunts is a very interesting one. Among the seafaring fraternity he is known as the goony. Their breeding places are confined to the southerd of 37°. Of course, they are seen up as far as 12° Lat. north, but very rarely. You will find rookeries on the Tristan da Cunha Island. These lie in Lat. 37° S., but you will never get them above 37°. They are to be meet with on all of the southern Antarctic islands. But where [they] reign supreme in their solitary and ancient haunts is in the Crozet group, Prince Edwards, Kerguelen Land, and clean down to the [south] Shetlands and South Georgia. They are also found in the Antipodian islands as far south as 40° S. in the Pacific Antarctic circle.

Their size is gigantic as they measure from wing to wing as much as fifteen and a half feet and when standing upright will measure two or three feet high. Of course, there are large ones and about . . . years old; the younger ones vary in size according to age. They have a white head in color, shaped something like a goose, only their bill is not so wide and about six inches long and inclined to curve down like a hawk's. They have medium sized eyes and black in color; neck and breast are of an alabaster white. They have a very thick tuft of feathers for a tail and about seven or eight inches in length. Underneath it is of a snow white color and on top it is of a slatish glossy black. The legs are very awkward looking ones for the bird. The feet are of the palmipede [that is, palmate] kind, very broad, and measure from six to eight inches across. The film that joins them is very thick, more so than other web-footed birds. It is of a light slatish color. The feet are adorned with three nails about one-third inch long.

They are not a timid bird when in their native haunts, but can be approached within a short distance before they will take flight. They always select a level place for their nests, with elevated ridges about it. Their idea in having their nests among ridges is for convenience in rising in the winds. They get on these ridges, face the wind, and then they run along a little distance and rise. Even when they rise off of the water, they will always rise on the crest of a wave. Their nests are like a pyramid with the top cut off in shape and are from a foot and a half to three feet in height and about three feet in circumference. They are composed of grass and moss. The latter material is very suitable for this purpose as it is very soft and quite long and of a yellowish color. They are very substantially constructed and they will spend several months making them. The old ones are remodeled every season. They are very clean about their nests and keep it continually in repairs.

They come in vast numbers from the sea on about the latter part of October. They are very awkward when they alight as they will at first tumble about on the ground and after taking in the surroundings will start and strut up to their nests, something after the fashion of a sailor who has been out on a long voyage, or [like a] goose. They then go up and look at their nests and [a] very queer thing is that they will never confound them. Then a pair of them will get together. Male and female rub their beaks together, clap their wings, and make a noise like the whinnying of a horse. They are very affectionate when they first arrive, and I suppose they recount to each other the different adventures that they have met with during their long flight across the seas. The female — I have found it an impossibility to tell which — settles down on the nest. But before taking possession of it they stay around the place, taking short trips to sea, in quest of food, I suppose. And about the first of December she settles down on the nest for good.

They lay but one egg and it puts in its appearance about the 26th of December. They take turns in setting on it. For instance, the male — but as I have said it is an utter impossibility to tell one from the other — will take his position on the nest and the female will go in quest of food. After an absence of about one day or more she returns and after a good deal [of] whinnying will relieve the [male], who will do the same thing. This keeps up and about the first of March the offspring puts in his appearance.

When first hatched, he or she — whichever it may be — is about as large as a good sized gosling, looking very much like a white ball of cotton or young gosling only a little larger. They are very cute looking and very playful in their ways. Both birds take turns in feeding him and treat him with extreme affection. They will go and return at intervals with a goodly supply of squid. The young one will present his open mouth to the parent birds and they will deposit the food in it. After the young commence to grow larger, this is about the 1st of June, the birds will go off on a series of voyages and remain away for a day and whenever they return they act very affectionate to each other.

Now they will commence to absent themselves for a longer period. They will both leave now and remain for a week or a fortnight. During this time, of course, the young one is learning to take care of himself. About the latter part [of] October, the parent bird abandons the nest for the season and [the] young bird is left to face the world himself. He first attempts to fly when he is about seven months old, and a very awkward attempt he makes. He will get up on one of these elevated ridges and commence to flap his wings valiantly. He will then take a run along the top of the elevation and vault away in the air. His first attempt won't prove of much avail to him, as he will fly a little distance and come down on the ground in all a heap. He will then pull himself together, looking much wiser for the experience of the first attempt. A very singular thing is that they never eat during the absence of the parent bird, until he leaves for good. You might say that he lives on his own fat for four months.

During the period after his hatching he grows very fat and plump, as he will weigh some fifty pounds. He subsists on this surplus fat, and by the time that he is about to leave, he has grown into a mere skeleton of his former self. He has a very thick layer of fat of a whitish color. During this time they don other feathers, the white ones falling off.

Eventually he becomes so he can fly a good distance. He will then take to the water. He is now able to take care of himself. He arises from the water and reels off again. He is not seen anymore ashore until he comes in next season to breed. While out at sea I suppose he falls in with a better half, and both cast lots together. The bird, when young, makes splendid eating and has a taste liken to veal. They can be killed very easily, especially when young and just out of the nest. The one precaution that you have to take in opening them is to take care and not burst the oil bag in them, for it will give the meat a very unpleasant taste and flavor.[23] You then skin them and taking care that you remove the fat from off the flesh, as it is better so, and if pork grease is a scarce commodity with you, you can use it to cook the flesh in. The old ones when attached or caught will emit an oilish substance from their mouth having a very strong smell. I suppose they do this in hopes that it will ward off the attacker. They are often caught at sea by the means of a fish hook, baited with pork and let tow astern. The best time to do this is when you are lying to in a gale of wind or in a dead calm.

BIRDS OF THE ALBATROSS FAMILY

Excepting albatross be mollymock, peale, and [nellie] or stinker, these three birds belong to the albatross family and are species of that bird.[24] Of the three, the mollymock resembles an albatross the most of the family. Of course the resemblance is only in color and shape but not in size or habits. This [bird] makes its nest where they are inaccessible to man, way up in the rugged cliffs where it is impossible to get without the aid of a rope and always near water. Their nests are composed of sticks and moss and they lay two or three eggs of a snow white color and the size of a chicken's eggs. They lay in October and the brood is ready for flight in May. Their young they are of a brownish color and as they grow older white predominates. Their similarity is so much like that of the albatross that at a distance the difference can't be told. They are very good eating. In the Cape Colony, South Africa, they are known as cape geese and are sold at the market like fowls. They are quite a source of revenue to the native fisherman. They catch [them] by the means of a hook baited with meat.

The peale:[25] this bird is about two-thirds the size of an albatross proper. Its plumage is of a coffee color. Only around the eyes they are adorned with small ringlets of white and very rarely you will see one with a white ring around his neck. Their beak is black and inclined to curve down at the end. They inhabit the inaccessible cliffs but do not congregate together in rookeries, as you hardly ever see but one or two pair together. Their nests are composed of moss [and] sticks, and they lay two or three eggs, white and of the size of a chicken's. They can be used for eating purposes.

BIRDS INHABITING THE ISLAND OF KERGUELEN

The snow bird (or paddy), sea hen, shag, the mackerel gull, stormy petrel, cape pigeon, Desolation teal duck, blue night hawk, stinker night hawk, and the diver. The above list comprises the birds that inhabit this island, excepting of course the albatross and birds belonging to that species and then penguins.

The snow bird:[26] it is known also as the paddy. This bird is about the size of a young chick. His plumage is snow white. It has a head similar to that of a pigeon, black eyes, and a black beak. This contrast gives the bird a very odd and comical appearance. Its beak is so constructed that it is very easy for the bird to open or get into any kind of crustaceous fish. Its wings are about the same as those of an ordinary pigeon, although they are not as extra good flyers. This bird being unamphibious, its feet are not of the palped [sic; palmated] sort. Its legs are covered over with feathers something after the fashion of those of a bantam rooster. They go in pairs and not in flocks, the year around. Their nests are constructed out of moss and grass and are built under cliffs. The eggs are the size of those of a pullet, white with brown specks.

He is quite tame and very inquisitive. If you happen to be seated and he sees you, he will come or fly down in front of you and commence to size you up. He will even go so far as to come close enough to take a peep into the barrel of your gun. Having satisfied his curiosity, he will go away. They are great thieves as they will steal eggs and eat them. I have seen the men getting eggs; they would collect a large pile of them and absent [themselves] getting more. And when they would return, to their dismay they would [find] that Mr. Paddy had helped himself to them by picking a hole in each one. Whether it is for mere mischievousness or for the sake of eating they do it, I do not know. This bird is fitted out with a gizzard pretty much the same as that of any fowl.

The sea hen:[27] this is a patalpeded [palmated] bird. Its color is of a coffee shade. They are about the size of a good sized hen. There is nothing remarkable in their appearance. They are very fleet of wing and [the] most vicious and defensible bird on the island. They build their nests amongst the grass, preferring a place thickly covered with grass. If you unknowingly approach their nest you will be notified of it in a jiffy, as they will boldly fly to you and pick at you and move to you. If there is any number of them, you are liable to come out minus an eye or some hair. I have frequently had them attack me when [I was going] after elephant. When the bird would swoop down at me, I would present my lance to it. Flying with such swiftness it would dash itself onto the lance, with the result that the lance would pass through its body and of course kill it. The food of this bird is squid, and I suppose fish of different kinds.

The shag is a water bird, a proper aquatic bird.[28] They have a long beak and a medium sized head. Their wings are not suited for flying, as the feathers are not very long, and not as elastic and pliable as those of other birds. Their backs are of a black velvety color; their belly is white as snow. They are excellent swimmers, being as fleet as a fish in the water. When young they are of a dark brown color and as they mature they turn black and white. Their feet vary in color. They are the only bird that inhabits these parts that is not fit to eat, their flesh having a very strong and rancid flavor, liken to that of old stale fish. They are constantly to be seen in the water, and if approached will disappear to other parts. They build their nests in cliffs and lay generally half a dozen snow white eggs. For curiosity's sake I have killed a number of them and in their entrails I have found stones of the size of a glass marble; I think they swallow these for ballast. They are never seen very far off-shore and it is a sure indication that you are nearing land when you see them.

The mackerel gull is a very pretty bird.[29] Their color is a slatish blue. [They are] about the size of a black bird. They have a red beak and a tuft of jet black feathers on their head. Their feet are also of a red flesh color. These bright and contrasted colors give the bird a very pretty appearance. They frequent the water a good deal, alighting on it to procure their food. They build their nests in low marshy places; generally a half a dozen or so will have the nests in close proximity. They lay one or two eggs, white in color with a blend of green. When approached they will set up an unearthly scream and if you do not retire they will dart at you in a very savage manner.

Desolation teal duck is the name given to a duck that inhabits these parts.[30] It differs but little in appearance to the bird at home. In fact, it is almost an exact counterpart. It has the characteristic steel green feathers in its wings and its body is of a greyish color. They are very plentiful here and are excellent eating. The eggs retain an earthy taste unfit to eat. They pair and build their nests in the low and marshy places around fresh water holes. During the mating season they are incapable of flying much. They lay half-a-dozen eggs and of a bluish white in color. Their food is principally the seed of the cabbage. That they can be domesticated has already been tested by me, and I find it a success. I have never brought any across the waters with me, but have had them aboard the vessel long enough to know that they will live in captivity. They become quite tame. And we had no more trouble with them than they have with most fowls.

The stormy petrel or Mother Currie's chicken:[31] some old salts actually believe that when a sailor dies his soul takes the form of one of these birds. They are seen most frequently at sea in a gale of wind or tempest, hence their name. The kind that inhabits these parts is a very small species, even smaller than our swallow. They resemble this bird very much and are often called sea swallows. The color of them is about the same, only their tail is not of the swallow kind. This bird lives by suction or by skimming along the water and picking its food up. A very small and minute jellylike fish is their principal food. There is an absurd belief that this kind carries its egg concealed on its body to sea with it. But this is the worst kind of cock-and-bull story. They make their nests on the sides of cliffs by excavating a small hole. Here they hatch and rear their young.

The diver:[32] this bird, next to the penguin, is given more to the water than any other bird that inhabits these parts. On the average they will weigh one-quarter pound. They are exceedingly fat and inclined to be very plump. Their wings are not fitted for flying as the feathers are very stubby and more quill than feather. They are excellent swimmers, being very fleet in the water. Their color is black on back and gray on the belly. Their feet are black and quite short.

APPENDIX 4

STORES USED ON KERGUELEN ISLAND

A. List of stores; Lawrence & Co. Papers, G. W. Blunt White Library, Mystic, Col. 25, box 8, folder 13. The totals, which do not add up, are given as on original document.

Captain & owners, schooner *Pilot's Bride*, to Lawrence & Co.
In amount below rendered $1,156.81
Interest from October 1881 to August 1884   206.69
                                                                          $1,363.50

Stores claimed to have been [taken] by Capt. Fuller:

4 casks bread, actual weight 2,232 lb.	133.92
18 barrels pork @$20	360.00
90 gallons molasses	27.00
1 foresail 266 yds	40.00
1 jib	25.00
1 flying jib	20.00
1 gaft [*sic*] topsail	25.00
1 staysail	35.00
1 royal	15.00
2 royal yards	10.00
	690.92
interest from October 1881 to August 1884	137.46
	$828.38

Additional claimed [by Lawrence & Co.]

5 barrels [?] beef	80.00
1 barrel hams	38.20
1 barrel sugar	30.20
1 barrel butter	21.50
1 barrel tongs [i.e., cod tongues]	12.00
1 barrel beans	12.50
1 barrel peas	12.50
60 pounds coffee	12.00
1 barrel dried apples	27.49

1 box pres[erved] corn	6.00
dried potatoes	6.00
1 barrel meal	6.00
	$264.39

1 mainsail	60.00
1 main staysail	30.00
1 main topgallant sail	30.00
	$180.00
	264.39
	$444.39

B. List calculated by Lawrence & Co., G. W. Blunt White Library, Mystic, VFM 235; source unknown, but apparently drawn up in August 1883 as Fuller was about to sail again for Desolation.

Joseph J. Fuller, of Danvers Mass., his Interest in Schr. *Francis Allyn* cargo & outfits for property taken at Pot Harbor Desolation from October 3rd 1881 to August 1882 without leave valued at $2,500, damages for taken same 2,500, and he he [*sic*] is about leaving the County.

List of Provisions and other Articles left at Pot Harbor on Sept. 10th 1881.

4 casks bread 2,232 lbs.
1 cask molasses 130 gallons
16 barrels [?] Pork
1 barrel peas
1 barrel hams
1 barrel sugar
1 barrel beans
1 barrel beans
6 barrels packet beef
1 box coffee
1 box corn
2 tins dryed potatoes
1 barrel dried apples
1 barrel butter
1 barrel meal
1 barrel cod tongues

1 cask containing
  1 mizen [sic] top staysail
  1 gaft topsail
  1 royal
  1 foresail & 1 jib
1 cask containing
  1 mainsail
  1 main staysail
  1 new main top. gal. staysail
1 cask [containing]
  1 main top. gal. sail
  2 royal
1 cask containing
  1 main top. gal. sail
  1 fore top. galsail, & rigging
All standing and running riggin [sic] belonging to two topgallant
  mast yards and royal mast and yards
upper cutting blocks
1 bellows 1 barrel coal
2 setts [sic] of whale Craft for 2 boats*
Jac'b ladder and manroops [sic: Jacob's Ladder and manropes]
1 barrel charcoal
3 top galt. yards
2 royal yards
2 top galt. mast & roy. mast.
1 boat mast & [?]
1 boat load of wood

[The following calculations, on an addition page, were probably an earlier reckoning of the same debt.]

To provisions & other articles taken at Pot Harbor, Desolation:

2,232 lb. bread	6¢	133.92
110 gallons molasses	30¢	33.00
16 barrels [?] pork	$20	320.00
6 barrels packed beef	$16	96.00
1 barrel ham 191 lbs	20¢	38.20
1 barrel sugar 302 lb.	10¢	30.20
1 barrel butter 86 lb.	25¢	21.50
1 barrel cod tongs [i.e., tongues]		10.00
1 barrel beans 5 bushels	2.50	12.50

*whale craft: harpoons and lances

1 barrel peas 5 bushels	2.50	12.50
1 box 60 lbs. coffee	20¢	12.00
1 barrel dried apples 229 lbs.	12¢	27.49
1 box corn (preserved)		6.00
2 tins dryed potatoes 30 lb.	20¢	6.00
1 barrel meal		4.50
one cask sails, containing mizzen topmast staysail, 165 yds.		33.00
1 gaft topsail $25, 1 royal $15		40.00
1 foresail 240 yds. $40, 1 jib $20		60.00
one cask containing 1 mainsail 316 yds., $60, 1 main staysail $20		80.00
1 main topgalt. staysail, new, 125 yds.		30.00
one cask containing 1 main topgal. sail $30, 2 royals $30		60.00
one cask [containing] 1 main topgal'tsail		30.00
1 fore topgal'sail		30.00
all standing & running rigging for 2 topgals., mast & royal mast; 2 topgal. yards, 2 royal yards, 2 topgal. mast. and royal mast		30.00
		$1,156.81
Interest from October 1881 to May 1883		92.48
		$1,249.29

NOTES

INTRODUCTION

1. Nathaniel W. Taylor, *Life on a Whaler, or Antarctic Adventures in the Isle of Desolation*, ed. Howard Palmer (New London, Ct.: New London County Historical Society, 1929, reprinted 1977); Rhys Richards and Helen Winslow, "The Journal of Erasmus Darwin Rogers, the First Man on Heard Island," *Turnbull Library Record* (Wellington, New Zealand), 4:1 (May 1971), pp. 31–43.

2. John Nunn, *Narrative of the Wreck of the "Favorite" on the Island of Desolation: Detailing the Adventures, Sufferings, and Privations of . . .* , ed. W. B. Clarke (London: Painter, 1850); for a similar account of the Crozets, see Charles Medyett Goodridge, *Narrative of a Voyage to the South Seas, and the Shipwreck of the "Princess of Wales" Cutter, with an account of Two Years Residence on an Uninhabited Island . . .* (Exeter, Eng.: Featherstone, 1844).

3. Raymond Rallier Du Baty, *15,000 Miles in a Ketch* (London: Nelson, [1912]), p. 325.

4. Frank C. Damon, *Reminiscences of the Linden Hill District of Danvers* (Salem, Mass., 1924); and with Florence A. Mudge, "The Romance of Joseph J. Fuller and Mary Ann Glass on the Island of Tristan da Cunha," *Historical Collections of the Danvers Historical Society* 30 (1942); 1–17; Harriet Silvester Tapley, *Chronicles of Danvers (Old Salem Village) Massachusetts, 1623–1923* (Danvers: Friends of the Peabody Institute . . . , 1923, reprinted 1974).

5. Damon and Mudge, op. cit.; W. Robert Foran, "Tristan da Cunha, Isles of Contentment," *National Geographic* 74 (November, 1938): 671–94.

6. Fuller family album of clippings, G. W. Blunt White Library, Mystic, Ct., Ms. 104; Wall collection, New London County Historical Society; *Cedar Grove Cemetery*, 1 (1936–41), 2 (1976).

7. Robert Owen Decker, *Whaling Industry of New London* (York, Pa., Liberty Cap Books, 1973), and *The Whaling City: A History of New London* (Chester, Ct., Pequot Press, 1976); Dale Plummer, "Henry P. Haven: Whaling Agent," *The Log of Mystic Seaport* 27,3 (November 1975), 66–73; Barnard L. Colby, *New London Whaling Captains* (Mystic, Ct.: Marine Historical Association, 1936).

8. New London Registry records, G. W. Blunt White Library, Mystic, Ct. The *Francis Allyn* was used as a steam sealer in the 1870s, but her boilers and machinery were removed in 1877, for, as *The Connecticut Gazette* of New London (13 July 1877) reported, "The application of steam to the *Allyn* was only an experiment and was found to be less profitable than expected." I am indebted to Bill Peterson for this reference.

9. Wreck Report of schooner *Pilot's Bride*, Federal Archives and Record Center, Waltham, Mass., Records of the Collection District of New London, Wreck Reports, 1881–1887, vol. 181.

10. New Bedford Registry records, Old Dartmouth Historical Society and Whaling Museum, New Bedford.

11. Whaleman's Shipping Paper, *Pilot's Bride*, Federal Archives and Record Center, Waltham, Records of Collection District of New London, Shipping Articles, Box 87b. The original crew list was lost with the vessel; Fuller had shipped additional hands in the Azores and Cape Verde Islands and one man in Cape Town, as described in the manuscript.

12. See for data of voyages, appendix 1 and sources given there. James Glass, cooper on the *Pilot's Bride*, kept a log of the 1887–1888 voyage of the *Francis Allyn;* see Ms. Log 190, G. W. Blunt White Library, Mystic, Ct.

13. Elmo Paul Hohman, *The American Whaleman: A Study of Life and Labor in the Whaling Industry* (New York: Longmans, Green, 1928), p. 13.

14. Data taken from the [U.S.] Defense Mapping Agency, Hydrographic Center, *Sailing Directions for the South Indian Ocean* (Washington, D.C., revised edition, 1975), p. 410 (figures rounded to nearest 1/10); further detailed information may be obtained from [U.S. Navy], Naval Weather Service Command, *U.S. Navy Marine Climatic Atlas of the World*, III, "Indian Ocean" (Washington, D.C., 1976).

PART I

1. To avoid later confusion, it should be pointed out that the articles did not stipulate the length of the voyage.

2. Fuller first sailed on the schooner *Franklin*, 119 tons, Edwin Church master, H. C. Chapell, agent. Church was a successful and well-known captain.

3. Left blank by Fuller. The Cape Verde Islands lie at 14°47′ West, 17°13′ North.

4. Trindade, a small volcanic island in the South Atlantic belonging to Brazil should not be confused with Trinidad in the West Indies.

5. Blackfish or Pilot Whale *(Globicephala melaena)*, normally under twenty feet in length, yield from five to eight barrels of oil.

6. In addition to the remarks in this section, Fuller describes the flora and fauna of Kerguelen in appendix 3.

7. The "after gang" included all the officers who lived aft. On a small sealing schooner, the boatsteerers, cooper, cook, and steward also lived aft. In larger vessels, they had separate quarters amidships.

8. The lance described is not to be confused with the longer whale lances used to kill the whales once taken by a whaleboat. The whale lance was normally longer (ten to twelve feet) overall, and the cutting blade shorter (three to four inches). Fuller distinguished clearly between the types in Part II (see below, p. 242).

9. "Rafting" and "backing" are described in more detail below, pp. 55–56 and 143–44. Similar descriptions may be found in Charles M. Scammon, *The Marine Mammals of the North-western Coast of North America . . .* (New York: Dover, 1968), pp. 118–23.

10. The nellie or stinker is a form of petrel; the sea hen is the Antarctic Skua. See pp. 313–14.

11. Actually, a fold in the skin, or "brood patch."

12. Mutton Cove cannot be precisely located.

13. Bark *Dove:* New London; 151 tons; Captain Smith; R. H. Chapell, agent.
14. Schooner *Exile:* New London; 83 tons; Capt. A. Tillinghast; E. V. Stoddard, agent.
15. As will be seen, whaling does not form a large part of Fuller's account. The three types of whale mentioned were the valuable and catchable baleen right whale *(Eubalaena glacialis);* the rather less desirable baleen humpback *Megaptera novaengliae);* and the largest of the toothed whales, the sperm whale *(Physeter catodon).*
16. Bark *Monticello:* New London; 353 tons; Captain Church (first name not known); R. H. Chapell, agent.
17. Typical ships' articles, in this case for the schooner *Pilot's Bride,* are included in appendix 2; the form was a printed one provided by the government, to which additions might be made.
18. Rodriguez Island, in the Mascarene group (a dependency of Mauritius), 500 miles east of Madagascar.
19. Bark *Milwood:* 254 tons, L. Gruninger, master; G. Allen & Son, agent; New Bedford; whaling in Indian Ocean, 1857–1861.
20. "Chaplin" cannot be identified; possibly Fuller means R. H. Chapell, or Herbert L. Crandell (1844–1927), a bookkeeper and later an executive of Williams & Haven (see pp. 276–77).
21. The Crozets, a group of five islands, lie in the Indian Ocean at 46°30' S., 51° E.
22. The *Roman* is given by Starbuck as a ship, 350 tons, Captain Swain, Williams, Haven & Co.; voyage to Heard Island, 1873–1874. See, Alexander Starbuck, *History of the American Whale Fishery* . . . (New York: Argosy-Antiquarian, 1964).
23. The report of the famous exploration ship H.M.S. *Challenger* — at 2000 tons, this steam corvette was rather bigger than the usual visitor to Kerguelen — fills fifty volumes. A recent, convenient summary by Eric Linklater, is *The Voyage of the Challenger* (Garden City, N.Y.: Doubleday, 1972). The most authoritative account is still the first volume of the official report.
24. Capt. (later Sir) George Nares; the *Challenger* carried several lieutenants, two of whom, Lord George Campbell and W. J. Spray, later wrote accounts of their adventures.
25. The transit of Venus between earth and sun (that is, across the face of the sun), occurs in cycles of 243 years at intervals of 121½, 8, 105½, and 8 years. Thus, in the nineteenth century, the transits occurred in 1874 and 1882, and will next occur in 2004. Observations of the transit were important in estimating the distance of the earth from the sun, and a variety of observation points were necessary to observe the precise time of entry or exit of Venus onto the sun's disk. See Harry Woolf, *The Transits of Venus: A Study of Eighteenth-Century Science* (Princeton, N.J.: Princeton University Press, 1959), for the importance of the event; details of the 1874 transit may be found in Richard A. Proctor, *Transits of Venus* . . . (New York: Worthington, 1875).
26. Starbuck gives the *Emma Jane* as a schooner of 86 tons, belonging to Williams, Haven & Co. However, he mentions a Captain Swain, without specifying his first name, as the master of the vessel.
27. Fuller's namesake is Port Fuller (or Fuller's Harbor), often called Little's

Harbor in Fuller's day. It lies on the southeast side of Howe's Island (69°28'29″ S.; 48°53'25″ E.) France, Commission Territoriale de Toponymie, *Toponymie des Australes*, comp. G. Delepine (Paris: Graphique, 1973), p. 152.

28. Exile Harbor cannot be precisely located.

29. Kerguélen-Trémarec did make two trips to the island; he did not commit suicide on the spot, but rather returned to write up his adventures. His *Relation de deux voyages dans les mers australes* was published in Paris in 1782. Kerguélen-Trémarec died in 1797.

30. Lest there be any doubt, Fuller was quite correct; meat, if it is fresh, plentiful, and not overcooked, is an effective antiscorbutic.

31. Fuller's account clearly is confused here; perhaps he means he was kept on board his vessel for another day by the weather.

32. H.M.S. *Volage* was assigned to tend the British transit of Venus party at Kerguelen. For further details, see Sir George B. Airy, ed., *Account of the Observations of the Transit of Venus, 1874* . . . (London: H.M.S.O., 1881).

33. Captain Nares's remarks on Fuller are found in the narrative volume of the *Challenger Reports* (vol. I, p. 346) and are conveniently quoted in Nathaniel W. Taylor, *Life on a Whaler* . . . (New London, Ct., New London Historical Society, 1929), p. 197. "Whaling operations [at Kerguelen] are confined almost entirely to the leeward (northeast) side; the weather side is, however occasionally visited by Captain Fuller in the *Roswell King*, who is thoroughly acquainted with the whole island. . . . Some idea of the danger of this enterprise may be formed from the fact that the *Roswell King*, a schooner of 100 tons, carries for use on the weather side of Kerguelen Island, an anchor and cable of the same size as that used in the *Challenger*, a vessel of 1420 tons." Many of the *Challenger* accounts refer to Fuller: for example, H. N. Moseley, *Notes by a Naturalist* . . . (London: Macmillan & Co., 1879), p. 186: "We fell in with an American whaling captain, Captain Fuller, who has been often on the weather shore. . . ."

34. Carpenter's Harbor cannot be precisely located; the same is true of a number of points on Royal Sound.

35. From this point of Part I onward, the ink is very faint in the manuscript. Fuller added a note of explanation: "Owing to the scarcity of ink aboard, I am obliged to make a mixture of vinegar, etc., etc. I can assure you that the congluberation would puzzle the very best chemist to analyse it, as all of the acids and corrosive sublimates that my medicine chest contains go to compose the ingredients of this ink. Thanks [also] to my chief mate, [who] managed to give me part of his ink."

36. Schooner *Charles Colgate*: New London; 250 tons; Capt. William Sisson; Lawrence & Co.

37. Rev. S. J. Percy was in charge of the British transit team on Kerguelen; his account may be found in Airy, op. cit.

38. A small section of manuscript is missing at this point.

39. Fuller might have added that ambergris is found only in sperm whales.

40. The U.S. transit of Venus party was landed in September of 1874, by the U.S.S. *Swatara* (3rd rate), establishing their observatory at Point Malloy on Royal Sound. The party consisted of Commander Ryan, and Lieutenant Commander Train, astronomers, and Dr. J. H. Kidder, surgeon, all U.S.N., and Messrs. Holmes, Dryer, and Stanley, photographers. In addition, they were

assisted by a cook, a carpenter, and "three boys, stowaways from Cape Town, afterward turned over to the British man-of-war." (J. H. Kidder, "Contributions . . . ," *Bulletin of the U.S. National Museum* 2[1875]: vii.) In mid-October a British party established itself one mile to the southwest, and about the same distance to the northwest was the German frigate *Gazelle*. The ships themselves soon departed. A month before the *Swatara* was expected to return, the American party was surprised to see the U.S.S. *Monongahela* arrive with orders to remove them — this on 9 December, the day of the transit. The *Monongahela* was persuaded to stay until 12 January.

41. The ship was the German frigate *Gazelle;* see n. 40.

42. Author's Note: "Before leaving that morning [I] noticed officers of the *Monongahela* looking at the sun. 14 March. Transit was taking place." The date is inexplicable, unless it was the date Fuller added the note, for by then both vessels had departed from Kerguelen.

43. The U.S.S. *Monongahela*, barkentine-rigged screw sloop-of-war (eleven guns), launched 1862, Capt. James S. Thornton. The *Monongahela*'s deck log shows that she spoke to the American schooner *Roswell King* on 9 December 1874 (the day of the transit). Deck log, U.S.S. *Monongahela*, v. 28, Navy and Old Army Branch, National Archives, Washington, D.C.

44. The term "wig" refers to the hair on the seal, but was used by the sealers to mean the male seal, while "clapmache," supposedly derived from the name of a Dutch sailor's cap, applied to the female. Fuller's common spelling was "whig" and "clapmacce." See Charles Medyett Goodridge, *Narrative of a Voyage* (Exeter, Eng.: Featherstone, 1844), p. 51.

45. Boot Leg Bill cannot be located.

46. Snug Harbor cannot be precisely located.

47. The *Corinthian*, 505 tons, a well-known ship belonging to Perkins & Smith of New London. Captain Slate sailed for Desolation on 23 September 1847 and returned on 26 June 1849 with 3,700 barrels of oil.

48. It is not clear which voyage of the *Exile* is meant.

49. It is obvious from the account that follows that Captain Rogers is regarded as master of both the *Emma Jane* and the *Roman*. This is explained by the fact that the *Emma Jane* was simply a tender, or auxiliary vessel, serving the larger *Roman;* the *Roman*'s master thus made the important decision for both vessels of whether to stay or to return home. The *Emma Jane*'s actual master would be expected to do as he was told in this regard.

50. According to the New London *Day* of 13 July 1935, Mrs. Rogers gave birth to a daughter at Mauritius, but when she embarked again upon the *Roman*, she contracted typhus and died at Bourbon (Reunion). Captain Rogers took care of the daughter, Allie, but she died in May of 1876. Benjamin Nelson Rogers, the *Roman*'s master at this time, should not be confused with his brother, Erasmus Darwin Rogers, another well-known captain. (To make it doubly confusing, both commanded the *Colgate* at different times.) Clipping in Fuller scrapbook, Ms. 104, G. W. Blunt White Library, Mystic, Ct.

51. The *Colgate* was engaged in loading the *Roswell King*'s oil from 30 December 1876 to 4 January 1877. One incident that Fuller has not reported is hinted at in the *Colgate*'s log for 26 December 1876 (at Three Island Harbor): "At 5 P.M. Captain Fuller com [*sic*] on board and reported the *King* on the Rock about ten

miles from this place." The next day the *Colgate* went to the *King*'s assistance, and both were back at Three Island Harbor on the 27th. Log of schooner *Charles Colgate* 1875–77, Lawrence & Co. Papers, log 112, G. W. Blunt White Library, Mystic, Ct.

52. The *Monongahela* log shows that the *Monongahela* left Three Island Harbor on 12 January under steam with a falling barometer; the wind shifted southwest to northeast. With a rough sea, the vessel was pitching deeply.

53. This section of the manuscript is very confused; Fuller has written over the name of each vessel with the other, windward with leeward, etc.: the version that seems to have been written in the heaviest ink has been used.

54. The *Monongahela* log recorded that a medical board removed Captain Thornton from duty on 6 February for "physical and mental disability." His gear was sent on board the *King* and the *Colgate* for passage home.

55. Author's note: "Boer — a native farmer of Africa, not necessarily a Hollander."

56. The details were left blank by Fuller. The end of the manuscript must have been written sometime after the voyages recounted (see the last paragraph in particular). Not only does Fuller refer to his own shipwreck in the voyage of 1880–1883, but no Zulu chief was exiled to Saint Helena until after Isandhlwana and the defeat of Cetewayo in 1879. Cetewayo's son, Dinizulu, was exiled to that island with two of his uncles, wives, and retinue in 1889, and it is very probably this chief who is meant — which would mean Fuller's account was completed after his last whaling voyage.

PART II

1. For a more elaborate description of the island by Fuller, see appendix 3.

2. Morgan Bay cannot be precisely located.

3. By "cask" Fuller means barrels that have been made and then broken down again to save space; it is not to be confused with "casks" holding six to eight (or, in the case of water, ten) barrels. The cooper's main task was to rebuild them as needed.

4. The average price of whale oil, per gallon, was 39¢ in 1879, but 51¢ in 1880. A barrel of oil would be worth roughly $15; a sealskin, $5. Williams, therefore, had in mind (see below) a cargo worth $32,500. It should be noted, for clarity's sake, that the discussion concerned *fur* seals, which had not been taken in any quantity from Kerguelen for roughly a half-century.

5. Lawrence and Co.'s bark *Trinity* (with Capt. John L. Williams as master) was large for Kerguelen at over 400 tons. New London registration figures, G. W. Blunt White Library, Mystic, Ct., give it at 419 and 417 tons; Robert Owen Decker, *Whaling Industry of New London* (York, Pa.: Liberty Cap Books, 1973), p. 160, gives it at 317 tons throughout, following Alexander Starbuck, *History of the American Whale Fishery* . . . (New York: Argosy-Antiquarian, 1964).

6. The *Roman* was normally listed as ship-, not bark-, rigged, 350 (or 380) tons; a Williams & Haven ship until sold away to New Bedford.

7. "M. Fuller" was Moses Fuller, Joseph's elder brother by four years.

8. The numbers are confusing. The *Bride*'s Whaleman's Shipping Paper listed seventeen men. Neither Alexander Shields, blacksmith, nor John Edwards, or-

dinary seaman, is mentioned in the narrative; one, at least, probably did not sail. Sailing with sixteen required ten Portuguese to make up Fuller's crew of twenty-six. Since he signed on four men at the Cape Verde Islands (see below), "making twenty-eight all told," he thus added eight at the Azores. Unfortunately for clarity, he was rescued at Desolation with twenty-two men, having lost four to death at the island and two to desertion at Cape Town (but he had signed on one man, Jack Reed, in that city). Total: twenty-nine, or twenty-three to be rescued.

9. "Doctor" was a commonly used nickname for the cook on merchant and whaling vessels.

10. "Shooks" are cask staves tied in bundles.

11. Thirty minutes, not hours, would be a long period underwater for any seal. Ten to thirteen minutes between breaths would be more probable. Thirty-six hours in the same mudhole, however, is a different matter.

12. "Hauling" or "hauling up" is a sealing term meaning the arrival of seal or sea elephant on their beaches, particularly for breeding.

13. Captain Robert Glass was Fuller's uncle; see p. 283.

14. Beaming boards (Fuller often spells it "beeming") were boards upon which the skins were spread and fixed, soaked in water, and then scraped of flesh and fat (see below).

15. Captain Smith, of course, drove a hard bargain: at $.05 a gallon, Fuller had offered between $1,500 and $1,600.

16. A useful discussion of the relationship between master, crew, and consul may be found in Elmo Paul Hohman, *The American Whaleman* . . . (New York: Longmans, Green, 1928), pp. 73–83. Fuller normally could have expected sympathy — but not necessarily effective assistance — from consuls whose income derived from fees.

17. "Broad arrow" was the symbol of British fiscal authorities.

18. James G. Blaine was speaker of the House from 1869 to 1875, unsuccessful candidate for the Republican presidential nomination in 1876 and 1880, and U.S. secretary of state in 1881.

19. The *Lizzie P. Simmons* was a small schooner (eighty-nine tons) of C. A. Williams & Co. Later she was commanded by the well-known New London master, J. W. Buddington.

20. Humpback whales *(Megaptera novaeangliae)* were unpopular, for although big enough, they were difficult to approach and inclined to sink when killed, and, when finally secured, they gave comparatively less oil.

21. Diana shore is not located.

22. Blue night hawks are a variety of petrel, living, like most of that family, in underground burrows. See appendix 3, pp. 314–16.

23. "Duff": a steamed or boiled pudding of flour and water, often enhanced with raisins and spices.

24. Mount Carmel cannot be precisely identified.

25. Sutter Bight cannot be identified; the reading of the name is questionable.

26. Rock of Dunder, Tumber Island, and Mutton Cove cannot be precisely located among the many islets in Royal Sound.

27. Lob hash, or lobscouse ('scouse): a hash, usually of sea biscuit, potatoes, onions, and chopped meat (i.e., salt pork).

28. George Manice was often referred to in the manuscript as "Tom" or "Tony," perhaps a nickname; this has been eliminated to avoid confusion.
29. Boothead cannot be identified.
30. The sea hen is a variety of skua; see appendix 3, pp. 314–15.
31. Mary Hester Brown is not identified; reading is questionable.
32. King Harbor cannot be located.
33. Atlas Island was probably Abbot or Abbot's Island; see map, p. xxvi.
34. "Slush" was waste fat or grease from the ship's galley.
35. Butler's Harbor cannot be located, but was probably Bayley Bay; see map, p. xxvi.
36. The U.S.S. *Marion*, steam sloop of war, was sent to Heard Island from the South Atlantic Station.
37. Planksheers are horizontal timbers forming the outer limit of the upper deck at the sides of the vessel.
38. Jug Keys cannot be located.
39. Herbert L. Crandell (1844–1927) rose from bookkeeper to vice president of Williams & Haven and C. A. Williams & Co.
40. Frank C. Damon's interest in the Fuller family was serious: see his article, "The Romance of Joseph J. Fuller . . . ," *Historical Collections of the Danvers Historical Society* 30 (1942): 1–17, first written in 1929, but not published until 1942.
41. C. A. Williams did not own a share in the *Allyn* until her 1883 registry, but T. W. Williams, H. P. and T. W. Haven, and R. H. Chapell all owned shares in her at various times.
42. H. R. Bond was a partner of William Williams, Jr. (1787–1870, older brother of T. W. Williams), who operated a separate whaling business in New London (Decker, *The Whaling City* . . . [Chester, Ct.: Pequot Press, 1976], pp. 78–9); Bond was another town notable, being trustee of the hospital and treasurer of the telegraph company among other positions.

APPENDICES

1. Kerguelen does not lie within the Antarctic Circle.
2. Although not numerous enough to be exploited commercially, fur seals were to be occasionally found during the period when Fuller was writing. N. W. Taylor *Life on a Whaler* . . . (New London, Ct.: New London County Historical Society, p. 71) saw none in his visit in the 1850s, but H. N. Moseley, a naturalist with the *Challenger*, killed two himself (*Notes of a Naturalist* . . . [London: Macmillan & Co., 1879], pp. 188–89) and recorded a whaling schooner that took nearly one hundred from some small islands off Howe's Foreland.
3. "Desolation tea," *Acaena ascendens* (J. H. Kidder, "Contributions . . . ," *Bulletin of the U.S. National Museum* 2[1875]:32, 2[1876]:23, giving *Acaena affinis*). An infusion of the leaves of this plant was regarded as an antiscorbutic and febrifuge.
4. "Desolation cabbage," *Pringlea antiscorbutica* (named in honor of Sir John Pringle, a student of scurvy), is peculiar to Prince Edward, Crozet, Kerguelen, and Heard islands. A cruciferous annual, unrelated to the domestic cabbage, it has no near ally. The root stalks, some two inches in diameter, often extending three or four feet along the ground are woody in texture and have the flavor of horseradish. At the extremity grows a large leaf that resembles garden cabbage;

the outer leaves are coarse, but the inner, compact leaves are edible, as Fuller points out. The cabbage served as essential fodder for the animals brought to Kerguelen by sealers in the heyday of the trade; depredation by rabbits, however, has resulted in severe destruction of the plant on Kerguelen.

Enjoyment apparently was in the opinion of the eater. Taylor (p. 84) ate it constantly, "cut up in its raw state and eaten with vinegar, or cooked by itself, fried in fat, or boiled with beef and pork." R. Rallier du Baty (*15,000 Miles in a Ketch* [London: Nelson, 1912], p. 178) could not force it down without boiling it twice, however, and C. M. Goodridge (*Narrative of a Voyage* . . . [Exeter, Eng.: Featherstone, 1884], p. 77) similarly found on the Crozets that it had to be boiled three or four hours. The difference might well have been in the amount of stalk each included.

5. Several species of grass, lichen, herbs, water plants, and the like are described by Kidder and Moseley (passim). A number of Fuller's references, however, such as the "big fish" or "blue flower," cannot be identified from the information given. Fuller was not a trained naturalist, and his lists are far from inclusive. Nevertheless, his interest and effort must be considered exceptional for his time and profession.

6. The wingless fly, *Calycopterix moseleyi*, is peculiar to Kerguelen; Moseley (p. 192) describes it as "simply a long-legged brown fly, with very minute rudimentary wings. It crawls about lazily on the cabbage, and lays its eggs in the moisture between the leaves, about the heart of the plant." Moseley describes several additional flies and gnats.

7. A number of fish, including *Mobulidae* (rays or "devil fish"), are found in the waters of Kerguelen.

8. Author's note: "At Hell's Gate there is quicksand; [I] have seen men drive penguins [into it] and they would instantly disappear."

9. Both theories can be correct. Waves of incredible magnitude (those of 120 feet are not impossible) have been experienced in these latitudes. Fossilized plant life, on the other hand, exists in abundance on Kerguelen, from the coal beds described by Fuller to "beds of fossil wood" discussed by Moseley, found in large trunklike masses "in various states of fossilization, some of it being comparatively soft, other specimens extremely hard" (p. 195).

10. *Grand Turk* was a famous name in American sailing history (see, for example, Robert E. Peabody, *The Log of the Grand Turks*, Boston: Houghton Mifflin, 1926). Fuller's reference, however, is not to the well-known privateer of the War of 1812, but to a whale ship (324 tons) that sailed from Dartmouth, Mass. and then from New Bedford, and that, according to A. Starbuck (*History of the American Whale Fishery* . . . [New York: Argosy-Antiquarian, 1964], p. 365) was condemned in 1843 and broken up.

11. Fuller added the height of 6,120 feet to the original manuscript. The contemporary American chart gives the height as 6,430 feet.

12. Dr. J. H. Kidder was attached to the American transit of Venus party; see Part I, n. 40.

13. Author's note: "At Christmas Harbor there is a volcanic crater; it is not in operation."

14. Goldsboro cannot be identified.

15. The four varieties of penguin found at Kerguelen and described by Fuller are:

	mean weight (pounds)	maximum weight (pounds)	body length (inches)
Aptenodytes patagonica (king)	33.0	104.5	37.5
Eudyptes chrysolaphus (macaroni)	9.3	——	27.5
Eudyptes chrestatus (rockhopper)	5.5	9.9	21.5
Pygoscelis papua papua (northern gentoo; Johnny or Johnnie)	13.7	19.0	32.0

16. The difference might have been merely semantics; the egg is cradled in a flap of skin, known as a brood patch, not in an actual pouch or pocket as Fuller implies.

17. Common roosting of the king penguin and the wandering albatross would be unusual, but not impossible; isolated pairs of albatross are capable of roosting among other birds.

18. The missing months are, respectively, October and November. The incubation period of rockhoppers is from thirty to forty-three days.

19. The sea elephant or elephant seal (the terms are interchangeable) is a form of seal (genus *Mirounga*) found in two species, Pacific coast of North America and Antarctic.

20. The sea leopard or leopard seal (interchangeable terms), *Hydrurga leptonyx*, is a seal that preys on its lesser relatives and birds — particularly on penguins; it could be taken for oil, but its defensive characteristics, and the small amount of oil produced from it, made the sea leopard unprofitable from the sealer's standpoint.

21. Wandering albatross, or *diomedea exulans*.

22. The message was found in North Fremantle, Australia; a dead albatross had a band around its neck, on which was punched in tin, "13 naufrages sont refugies sur les iles crozet 4 aout 1887," the Crozets lying 3,500 miles away. By the time a relief expedition arrived, the party had moved to another island of the group in search of food and was not heard of again; William Jameson, *The Wandering Albatross* (New York: Morrow, 1959), p. 34. That Fuller was familiar with the incident shows that this section of manuscript was added after 1887.

23. The oil secretion of albatrosses and petrels is still unexplained. The oil occurs in large quantities in the stomach, is rich in vitamins A and D (but low in protein), turns to wax when cold (resembling spermaceti), and has an unpleasant, persistent, strong musky odor. It is alternately described being used by the birds for (1) feather preening; (2) infant feeding; (3) possibly, as a water supply for infant birds; and (4) protection, particularly since the stomach contents are jettisoned in preparation for emergency flight. The birds — particularly the fulmar — are capable of shooting the fluid several feet toward the intruder (perhaps less deliberately than because the birds face intruders). These various usages are not mutually exclusive. Jameson, pp. 60–61; R. M. Lockley, *Ocean Wanderers: The Migratory Sea Birds of the World* (Newton Abbot: David & Charles, 1974), pp. 29–30.

24. It is not possible to identify Fuller's "Birds of the Albatross Family" with precision. "Mollymock" ("mollymauk," "mollymawk," among other spellings) was a general term used by seamen for several albatrosses and petrels. The same was true of the terms "nellie" and "stinker," the latter obviously referring to the stomach oil. The terms were more particularly applied, however, to the giant

petrel *(Macronectes giganteus halli),* by far the largest of the species. Kidder (2[1875] 25–26) gives another petrel *(Procellaria aequinoctialis,* or "white-chinned petrel") as the "stinker" on the authority of the whalers of the schooner *Emma Jane.* Kidder added that "Captain Fuller, however, of the schooner *Roswell King,* a very careful observer, tells me that the stinker is a much larger bird, and that it nests on the ridges of the high hills, not in burrows, and very late in the season."

25. The term "peale" is less clear. Peale's petrel *(Pterodroma inexpectata* among other designations) is a bird of the Pacific, known as the "rain-bird" in New Zealand. Kidder (2[1875]) also explained, however, that the sooty albatross *(Phoebetria fusca)* was known as "pee-arr" to the sealers, as this was perhaps Fuller's intent. On these and the birds described below, I have followed W. B. Alexander, *Birds of the Ocean* (New York: Putnam, 1928).

26. The snow bird, or paddy, is the sheathbill, *Chionis minor,* also called the sea or white pigeon.

27. The sea hen is a skua, several species of which roost on Kerguelen (Alexander, pp. 203–13). The skua resembles the eagle or buzzard: it is essentially a bird of prey and feeds on carrion.

28. "Shag" is a common term for cormorant, again several species of which are found on Kerguelen, particularly the *Phalacrocorax verrucosus,* or Kerguelen cormorant.

29. "Mackerel gull" probably refers to *Larus dominicanus,* the southern black-backed gull.

30. Desolation teal duck *(Querquedula eatoni)* is a brown teal variant peculiar to Kerguelen.

31. Several varieties of stormy petrels are found at Kerguelen. They are known to seamen as Mother Carey's (or Currie's) chickens; "Mother Carey" is a corruption of Mater Cara, an appelation of the Virgin Mary. "Night hawk" refers to several smaller varieties of petrel, including the blue petrel (probably Fuller's blue night hawk), the brown petrel, and the white-headed petrel among others. The pintado petrel *(Daption capense,* or "Cape pigion") is very common in the southern hemisphere. It is distinguished by its checkered upper parts.

32. The Kerguelen diving petrel *(Pterodroma brevirostris)* was called "diver"; Kidder, 3(1876):36.

SELECTED BIBLIOGRAPHY

Airy, Sr George B., ed. *Account of the Observations of the Transit of Venus, 1874.* . . . London: H.M.S.O., 1881.

Alexander, W. B. *Birds of the Ocean: A Handbook for Voyagers.* New York: Putnam, 1928.

Budd, G. M. "The King penguin Aptenodytes patagonica at Heard Island." In *The Biology of Penguins,* edited by B. Stonehouse, pp. 337–53. Baltimore: University Park Press, 1975.

Cedar Grove Cemetery. New London, Ct.: New London Cemetery Assoc., I (1936–41), II (1976).

[Challenger, H.M.S.] *The Report of the Scientific Results of the Exploring Voyage of H.M.S. Challenger During the Years 1873–1876.* 50 Vols. Edinburgh: Challenger Office, 1885–1895. "Narrative of the Voyage," edited by J. Y. Buchanan, H. N. Moseley, J. Murray, and T. H. Tizard, I, parts 1 and 2.

Clark, A. Foward. "The Antarctic Fur-Seal and Sea-Elephant Industries." In *The Fisheries and Fishery Industries of the United States* edited by George Brown Goode, 2: 400–67. Washington, D.C.: G.P.O., 1887.

Colby, Barnard L. *New London Whaling Captains.* Mystic: Marine Historical Association, 1936.

Damon, Frank C. *Reminiscences of the Linden Hill District of Danvers, Reprinted from the Salem "Evening News."* Salem, Mass.: Newcomb & Gauss, 1924.

——, and Mudge, Florence A. "The Romance of Joseph J. Fuller and Mary Ann Glass on the Island of Tristan da Cunha." *Historical Collections of the Danvers Historical Society* 30 (1942): 1–17.

Decker, Robert Owen. *The Whaling City: A History of New London.* Chester, Ct.: Pequot Press (for New London County Historical Society), 1976.

——. *Whaling Industry of New London.* York, Pa.: Liberty Cap Books, 1973.

Foran, W. Robert. "Tristan da Cunha, Isles of Contentment." *National Geographic* 74, 5: 671–94.

France. Commission Territoriale de Toponymie. *Toponymie des Australes,* compiled by G. Delepine. Paris: Graphique, 1973.

Goodridge, Charles Medyett. *Narrative of a Voyage to the South Seas, and the Shipwreck of the "Princess of Wales" Cutter, with an Account of Two Years Residence on an Uninhabited Island.* 5th ed. Exeter, England: W. C. Featherstone, 1844 [1843].

Hegerty, Reginald B. *Returns of Whaling Vessels Sailing from American Ports: A continuation of Alexander Starbuck's "History of the American Whale Fishery."* New Bedford: Old Dartmouth Historical Society and Whaling Museum, 1959.

Hohman, Elmo Paul. *The American Whaleman: A Study of Life and Labor in the Whaling Industry.* New York: Longmans, Green, 1928.

Jameson, William. *The Wandering Albatross.* New York: Morrow, 1959.

Kidder, J. H. "Contributions to the Natural History of Kerguelen Island, Made in Connection with the United States Transit of Venus Expedition, 1874–75." *Bulletin of the United States National Museum* (Washington, D.C.: G.P.O.) 2 (1875), 3 (1876).

Linklater, Eric. *The Voyage of the "Challenger."* Garden City, N.Y.: Doubleday, 1972.

Lockley, R. M. *Ocean Wanderers: The Migratory Sea Birds of the World.* Newton Abbot: David & Charles, 1974.

MacKay, Margaret. *Angry Island: The Story of Tristan da Cunha (1506–1963).* Chicago: Rand McNally, 1963.

Matthews, L. Harrison. *Sea Elephant: The Life and Death of the Elephant Seal.* London: MacGibbon & Kee, 1952.

Migot, Andre. *The Lonely South.* Translated by Richard Graves. London: Hart-Davis, 1956.

Moseley, H. N. *Notes by a Naturalist on the "Challenger," Being an Account of Various Observations Made during the Voyage of H.M.S. "Challenger" Round the World in the Years 1872–1876* . . . London: Macmillan & Co., 1879.

Murphy, Robert Cushman. *Logbook for Grace: Whaling Brig "Daisy," 1912–1913.* New York: Macmillan & Co., 1947.

Nougier, Jacques. *Contribution l'étude géologique et géomorphologique des Îles Kerguélen.* 2 vols. Paris: Comité national français des recherches antarctiques, 1970.

Nunn, John. *Narrative of the Wreck of the "Favorite" on the Island of Desolation: Detailing the Adventures, Sufferings, and Privations of* . . . London: W. E. Painter, 1850.

Plummer, Dale. "Henry P. Haven: Whaling Agent," *The Log of Mystic Seaport* 27, 3:66–73.

Proctor, Richard A. *Transits of Venus.* New York: Worthington, 1875.

Rallier du Baty, Raymond. *15,000 Miles in a Ketch.* London: Nelson, [1912].

Richards, Rhys, and Winslow, Helen. "The Journal of Erasmus Darwin

Rogers, the First Man on Heard Island," *Turnbull Library Record* (Wellington, New Zealand), n.s. 4, 1:31–43.

Scammon, Charles M. *The Marine Mammals of the North-western Coast of North America, Described and Illustrated, Together with an Account of the American Whale-Fishery.* San Francisco: J. H. Carmany, 1874; reprinted New York: Dover, 1968.

Spray, W. J. J. *The Cruise of H.M.S. "Challenger." Voyages over Many Seas, Scenes in Many Lands.* London: Sampson, Low, Marston, Searle, & Livingston, 1876.

Starbuck, Alexander. *History of the American Whale Fishery from its Earliest Inception to the Year 1876.* New York: Argosy-Antiquarian, 1964 [1878].

Stonehouse, Bernard, ed. *The Biology of Penguins.* Baltimore: University Park Press, 1975.

Tapley, Harriet Silvester. *Chronicles of Danvers (Old Salem Village), Massachusetts, 1632–1923.* Danvers: Friends of the Peabody Institute Library of Danvers and Danvers Historical Society, 1974 [1923].

Taylor, Nathaniel W. *Life on a Whaler, or Antarctic Adventures in the Isle of Desolation.* New London, Ct.: New London County Historical Society (Occasional Publications, II), 1929 (photo reprint, 1977).

Tilman, H. W. *Mischief among the Penguins.* London: Hart-Davis, 1961.

United States. Navy. Naval Weather Service Detachment. *U.S. Navy Marine Climatic Atlas of the World.* III: "Indian Ocean." Washington, D.C.: G.P.O., 1976.

United States. Works Progress Administration (Massachusetts). Federal Writers' Project. *Whaling Masters. American Guide Series.* New Bedford: Old Dartmouth Historical Society, 1938.

Woolf, Harry. *The Transits of Venus: A Study of Eighteenth-Century Science.* Princeton: Princeton University Press, 1959.

INDEX